14th International Conference on Transport in Interacting Disordered Systems 2011

(TIDS-14)

Journal of Physics: Conference Series Volume 376

Acre, Israel
5-8 September 2011

ISBN: 978-1-62276-810-3
ISSN: 1742-6588

Printed from e-media with permission by:

Curran Associates, Inc.
57 Morehouse Lane
Red Hook, NY 12571

Some format issues inherent in the e-media version may also appear in this print version.

Printed by Curran Associates, Inc. (2013)

For permission requests, please contact the Institute of Physics
at the address below.

Institute of Physics
Dirac House, Temple Back
Bristol BS1 6BE UK

Phone: 44 1 17 929 7481
Fax: 44 1 17 920 0979

techtracking@iop.org

Additional copies of this publication are available from:

Curran Associates, Inc.
57 Morehouse Lane
Red Hook, NY 12571 USA
Phone: 845-758-0400
Fax: 845-758-2634
Email: curran@proceedings.com
Web: www.proceedings.com

The Superconductor-Insulator transition and metal insulator transition

012001 Emergent granularity and pseudogap near the superconductor-insulator transition N Trivedi, Y L Loh, K Bouadim and M Randeria 1

012002 Cooper pair islanding model of insulating nanohoneycomb films S M Hollen and J M Valles Jr. 12

012003 Wonderful life at weak Coulomb interaction: increasing of superconducting/superfluid transition temperature by disorder V E Kravtsov 22

012004 Comparative Study of the Magnetic Susceptibility of n-Ge near the Insulator–Metal Phase Transition by Means of ESR and SQUID A I Veinger, A G Zabrodskii, T L Makarova, T V Tisnek, S I Goloshchapov and P V Semenikhin 34

Electron Glass and Coulomb glass

012005 Ageing and relaxation times in disordered insulators T Grenet, J Delahaye and M C Cheynet 40

012006 Quantum Coulomb gap M Pino, A M Somoza and M Ortuño 48

012007 Numerical studies of relaxation in Electron Glasses M Ortuño and A M Somoza 55

012008 Screening and conductance relaxations in insulating granular aluminium thin films J Delahaye and T Grenet 62

012009 Elementary excitations and avalanches in the Coulomb glass M Palassini and M Goethe 68

Hopping conductivity

012010 Hopping Conductivity Enhanced by Microwave Radiation Z Ovadyahu 74

012011 Electric field dependence of charge-carrier hopping transport at large carrier concentrations in disordered organic solids: Meyer-Neldel and Gill energies I I Fishchuk, A Kadashchuk, M Ullah, H Sitter, N S Sariciftci and H Bässler 80

Anderson Localization

012012 Extension of Anderson's Localization Theory to Interacting Systems M Pollak 86

012013 Anderson localization in atomic mixtures K Ziegler and O Fialko 91

012014 The destruction of the virtual Anderson transition in a narrow impurity band of doped quantum well structures N V Agrinskaya, V I Kozub and D S Poloskin 97

Magnetotransport

012015 Angular and Temperature-related Specific Features of the Interaction between Light and Heavy Holes in Weakly Doped p-Ge:Ga at Low Temperatures A I Veinger, A G Zabrodskii, T V Tisnek and S I Goloshchapov 103

012016 Magnetoresistance in two-dimensional array of Ge/Si quantum dots N P Stepina, E S Koptev, A G Pogosov, A V Dvurechenskii, A I Nikiforov and E Yu Zhdanov 109

012017 Interference mechanism of magnetoresistance in variable range hopping: the effect of paramagnetic electron spins A V Shumilin and V I Kozub 115

Low dimensional systems and heterostructures

012018 Disorder and Interaction Effects in Quantum Wires L W Smith, K J Thomas, M Pepper, D A Ritchie, I Farrer, J P Griffiths and G A C Jones 121

012019 Fano-Kondo oscillations of the conductance and thermopower in a mesoscopic transistor A Aldea, M țolea and I V Dinu 128

012020 Distribution of the Entanglement Entropy in Disordered One-Dimensional Fermions R Berkovits 133

012021 Compositional disorder anomalies in Ga(N,P,As)/GaP quantum well structures K Jandieri, S D Baranovskii, B Kunert, M Zimprich, S Liebich, K Volz, W Stolz, S Chatterjee, C Karcher, W Heimbrodt and F Gebhard 137

012022 Calculation of charge carrier concentration profile and magnetoresistance in modulation doped structures with a wide potential well L Shchurova 143

012023 Resonant Tunneling in a Polymer with Charge Injection Doping A Ionov and R Rentzsch 149

Quantum coherence, interference and dephasing

012024 Interference quantum correction to conductivity of $Al_xGa_{1-x}As/GaAs$ double quantum well heterostructures near the balance A V Germanenko, G M Minkov, O E Rut and A A Sherstobitov 157

012025 Quantum Coherence in Photosynthetic Exciton Dynamics P Nalbach and M Thorwart 163

14th International Conference on Transport in Interacting Disordered Systems (TIDS-14)

Aviad Frydman

Abstract Metrics

PREFACE

The '14th Transport in interacting disordered systems – TIDS14' conference took place during 5–8 September 2011 in Acre Israel. The conference was a continuation of the biennial meeting traditionally called HRP (hopping and related phenomena) and later named TIDS (transport in interacting disordered systems). Previous conferences took place in Trieste (1985), Bratislava (1987), Chapel Hill (1989), Marburg (1991), Glasgow (1993), Jerusalem (1995), Rackeve (1997), Murcia (1999), Shefayim (2001), Trieste (2003), Egmond, aan Zee (2005), Marburg (2007) and Rackeve (2009).

Central to these conferences are systems that are characterized by a large degree of disorder and hence they lack translational symmetry. In such systems interactions are usually very important. Dramatic differences in the behavior of crystalline solids and the 'disordered' systems are possible. Some examples of the latter are amorphous materials, polymer aggregates, materials whose properties are governed by impurities, granular systems and biological systems. This conference series is notable for the pleasant atmosphere and fruitful exchange of ideas between theoreticians and experimentalists in these areas. This tradition was also maintained in the conference in Israel.

Specific topics of TIDS14 included: hopping, electron and Coulomb glasses, Anderson localization and many body localization, noise, magneto-transport, metal–insulator and superconductor–insulator transition, transport through low dimensional and nanostructures, quantum coherence, interference and dephasing and other related topics.

Over sixty scientists from fourteen countries participated in the conference and presented papers either as oral presentations or as posters in two sessions that took place during the conference. Many of these papers are included in these proceedings.

I would like to thank all the conference participants for the interesting presentations, debates and discussions that created a stimulating but pleasant environment. Also, I thank the other members of the local organizing committee Tal Havdala and Shachaf Poran from Bar Ilan University and the members of the international advisory committee. I gratefully acknowledge support from the Israel Science foundation and from Bar Ilan University.

Aviad Frydman
Conference chairman
Bar Ilan University

Peer review statement

Abstract Metrics

All papers published in this volume of *Journal of Physics: Conference Series* have been peer reviewed through processes administered by the proceedings Editors. Reviews were conducted by expert referees to the professional and scientific standards expected of a proceedings journal published by IOP Publishing.

Emergent granularity and pseudogap near the superconductor-insulator transition

Nandini Trivedi[1], Yen Lee Loh[2], Karim Bouadim[3], and Mohit Randeria[1]

[1]Department of Physics, The Ohio State University, Columbus, OH 43210, USA
[2]Department of Physics and Astrophysics, University of North Dakota, Grand Forks, ND 58202, USA
[3]Institute for Theoretical Physics III, University of Stuttgart, Germany

E-mail: trivedi.15@osu.edu

Abstract. In two dimensions there is a direct superconductor-to-insulator quantum phase transition driven by increasing disorder. We elucidate, using a combination of inhomogeneous mean field theory and quantum Monte Carlo techniques, the nature of the phases and the mechanism of the transition. We make several testable predictions specifically for local spectroscopic probes. With increasing disorder, the system forms superconducting blobs on the scale of the coherence length embedded in an insulating matrix. In the superconducting state, the phases on the different blobs are coherent across the system whereas in the insulator long range phase coherence is disrupted by quantum fluctuations. As a consequence of this emergent granularity, we show that the single-particle energy gap in the density of states survives across the transition, but coherence peaks exist only in the superconductor. A characteristic pseudogap persists above the critical disorder and critical temperature, in contrast to conventional theories. Surprisingly, the insulator has a two-particle gap scale that vanishes at the SIT, despite a robust single-particle gap.

1. Introduction

A superconductor (SC) is an emergent state of matter in which electrons pair up forming Cooper pairs, the different Cooper pairs become phase coherent, and the system undergoes Bose-Einstein condensation. What is the effect of disorder on such a phase-coherent state? It was argued by Anderson [1] that three-dimensional superconductivity is quite robust, persisting even in polycrystalline or amorphous materials. Two dimensions turns out to be particularly intriguing because it is the marginal dimension for localization and superconductivity. One can ask: does a two-dimensional system develop superconducting behavior when all its eigenstates are localized due to a random potential and if so, what is the mechanism [2]? It is seen from experiments that superconductivity in two dimensions does exist but can be destroyed by a large variety of tuning parameters including temperature, inverse thickness (characterized by sheet resistance), disorder, gate voltage, Coulomb blockade, perpendicular magnetic field, and parallel magnetic field [3, 4, 5, 6, 7, 8, 9, 10, 11, 12, 13, 14, 15]. Once superconductivity is destroyed the system becomes an insulator beyond a quantum phase transition at $\epsilon_F\tau \approx 1$ (see Fig. 1). Much progress has made in the field of superconductor-insulator transitions (SIT) over the years, however, there are still several open questions:

Published under licence by IOP Publishing Ltd

- For zero disorder, we know that above the superconducting transition temperature T_c the system is a normal Fermi liquid. What is the nature of the state above T_c at finite disorder? Is it a Fermi liquid, or a quantum critical phase with a pseudogap?
- What is the nature of the insulator? Is it a localized Anderson insulator, a Mott insulator, a Fermi glass, a Bose glass, or something else?
- What are the energy scale(s) in the insulator that vanish at the SIT?
- What is the mechanism that drives the SIT?
- How do the single-particle spectral functions and dynamical conductivity behave in the superconductor, the insulator, and near the SIT?
- What is the universality class (critical exponents, amplitude ratios, and scaling functions)?
- Is the critical resistance universal and equal to the quantum of resistance $R_Q = h/(2e)^2$? If so (as in Ref. [15]), does this indicate self-duality?
- Can one develop a theory that captures both Coulomb amplitude suppression and phase fluctuations, and thus unify the fermionic and bosonic pictures of the SIT?
- Does the SIT happen concomitantly with the metal-insulator transition (MIT), or not?
- What is the origin of the gigantic peak in the magnetoresistance in the insulating state?

In our work so far we have successfully addressed the first five points and we summarize those in this article.

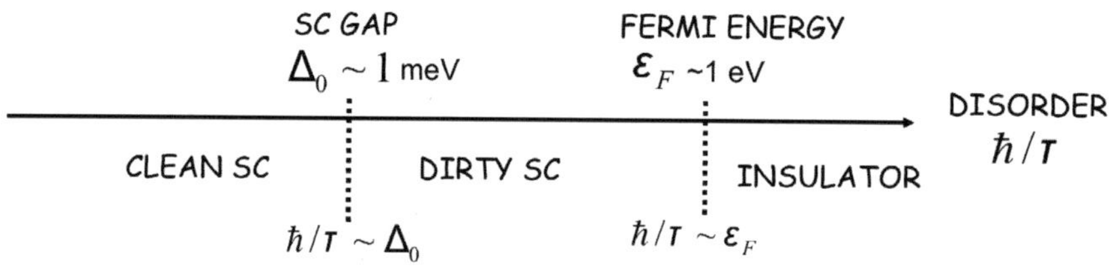

Figure 1. Typical Energy Scales: A superconductor has a gap of order $\Delta_0 \approx 1\,\text{meV}$, which is much smaller than the Fermi energy $\epsilon_F \approx 1\,\text{eV}$. Disorder introduces an energy scale $\hbar/\tau$ where τ is the mean scattering time. As the disorder is increased τ gets smaller and the disorder energy scale increases. When $\hbar/\tau \approx \Delta_0$ the system evolves from a clean superconductor to a "dirty superconductor" but superconductivity remains robust. It is only when $\hbar/\tau$ approaches the Fermi energy that the systems starts to show the superconductor to insulator phase transition.

2. Quantum Phase Transitions

A SIT is an excellent example of a continuous quantum phase transition (QPT) driven by quantum zero-point fluctuations controlled by a parameter V. For $V < V_c$ the system is a superconductor with well-defined Bogoliubov quasiparticles and collective modes of a superconductor, whereas for $V > V_c$ the system is an insulator, also with well-defined excitations. On the superconducting side, there is an energy scale that vanishes at V_c – the 2D superfluid density or phase stiffness. Similarly, on the insulating side, one generally expects an energy scale that vanishes at V_c.

The QPT also profoundly affects behavior of the system at finite temperature. The superconducting transition temperature T_c decreases with increasing V, finally vanishing at the quantum critical point (QCP) at $V = V_c$, as illustrated schematically in Fig. 2. The fan emanating from the QCP is a quantum critical region that is neither superconducting nor insulating, where the spectrum is broad and quasiparticles are ill-defined. The primary aim of experimental and theoretical research in this field is to clearly elucidate the properties of the phases and excitations, and then go on to understand the quantum critical region.

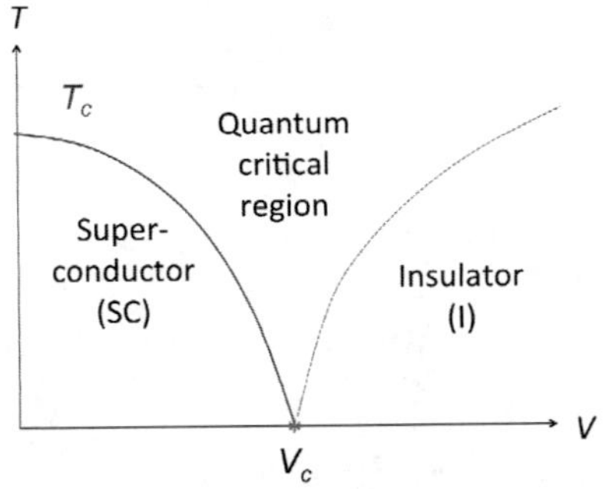

Figure 2. The SIT is a quantum phase transition occurring at zero temperature ($T = 0$) and at a critical value of a tuning parameter $V = V_c$. Above the quantum critical point there is a fan-shaped quantum critical (QC) region where fluctuations are enhanced and QC scaling is obeyed.

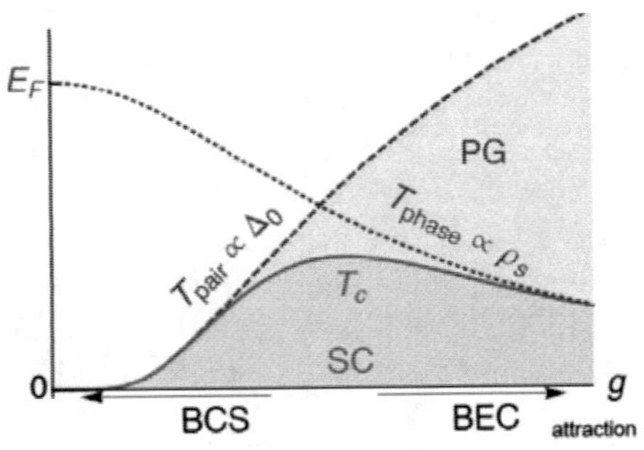

Figure 3. BCS-BEC crossover showing the energy and temperature scales of a generic 2D superconductor as a function of attractive interaction g.

3. Amplitude and Phase Fluctuations

In order to understand the role of amplitude and phase fluctuations, it is useful to consider a clean problem of fermions with attractive interactions. A singlet s-wave superconductor is described by a complex order parameter $\Delta(\mathbf{R}) = |\Delta(\mathbf{R})|\, e^{i\theta(\mathbf{R})}$. At zero temperature the pairing amplitude $|\Delta(\mathbf{R})|$ takes a uniform value, Δ_0. This is the energy scale associated with pairing. It typically manifests itself as an energy gap $E_g = \Delta_0$, and it also sets the maximum temperature, $T_{\text{pair}} = 0.57\Delta_0$, for the formation of Cooper pairs. On the other hand, the fluctuations of the phase, $\theta(\mathbf{R})$, are controlled by the superfluid density (or phase stiffness) ρ_s. For 2D superconductors ρ_s has the dimensions of energy, and it can be directly interpreted as the energy scale for phase fluctuations. ρ_s can be measured using mutual inductance techniques. It also sets the maximum temperature, T_{phase}, for long-range phase coherence.

As seen in Fig. 3, with increasing attraction the pairing scale $T_{\text{pair}} \propto \Delta_0$ increases, whereas the phase coherence scale $T_{\text{phase}} \propto \rho_s$ decreases (because tightly bound pairs have a larger effective mass in a lattice). The transition temperature T_c is determined by the lower of the two scales T_{pair} and T_{phase}. Weak-coupling superconductors such as Al are described by the BCS (Bardeen-Cooper-Schrieffer) limit of large overlapping Cooper pairs, where $T_{\text{pair}} \ll T_{\text{phase}}$. The phase coherence scale is of the order of the Fermi energy, $T_{\text{phase}} \sim E_F \sim 10^4\,$K, whereas the pairing scale is exponentially suppressed, $T_{\text{pair}} \sim 1\,$K. Thus, the critical temperature is determined by pairing, and we have a mean-field like, amplitude-driven transition. Conversely, in the BEC (Bose-Einstein condensation) limit where $T_{\text{pair}} \gg T_{\text{phase}}$, the Cooper pairs are small and tightly bound, and the transition is determined by the temperature T_{phase} above which phase coherence is lost. However, a pseudogap – a suppression in the density of states around the Fermi energy – persists up to a higher temperature T_{pair} [16, 17, 18, 19].

Although the crossover from BCS to BEC regimes is not directly relevant to superconductivity in weakly coupled traditional superconductors, we will see that as disorder is introduced, the amplitude-driven mechanism for the loss of superconductivity is modified to a phase-driven mechanism opening up a pseudogap regime. This manifests itself as the discrepancy between transport and tunneling measurements (Fig. 14): transport measurements are sensitive to global phase coherence and hence to T_{phase}, whereas tunneling densities of states are sensitive to local pairing correlations and hence to T_{pair}.

The key reason for the formation of a pseudogap regime even in a weakly coupled BCS superconductor is the phenomenon of disorder-induced "emergent granularity": a "homogeneously disordered" model or sample may have a strongly inhomogeneous pairing amplitude. This emerged from Bogoliubov-de Gennes (BdG) calculations [20, 21] and is supported by QMC work and experiments. See Figs. 4,6,7, and 8. At weak disorder the pairing amplitude is homogeneous. At strong disorder it breaks up into blobs. The phases of the order parameter on different blobs are weakly coupled, so the system is similar to a granular superconductor or a Josephson Junction Array (JJA), where phase fluctuations are extremely important (see Fig. 5). A treatment of phase fluctuations requires going beyond BdG.

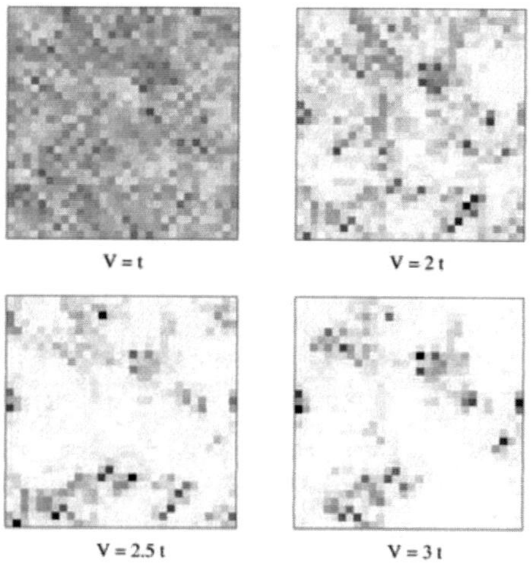

Figure 4. Maps of the pairing amplitude $\Delta(\mathbf{R}) = \langle c_{\mathbf{R}\downarrow} c_{\mathbf{R}\uparrow} \rangle$ for a given disorder realization with increasing disorder strength V, according to BdG calculations. For low disorder $V = 1t$ the pairing amplitude is fairly uniform over the system. With increasing disorder the system breaks up into SC puddles with large Δ in an insulating sea with $\Delta \approx 0$.

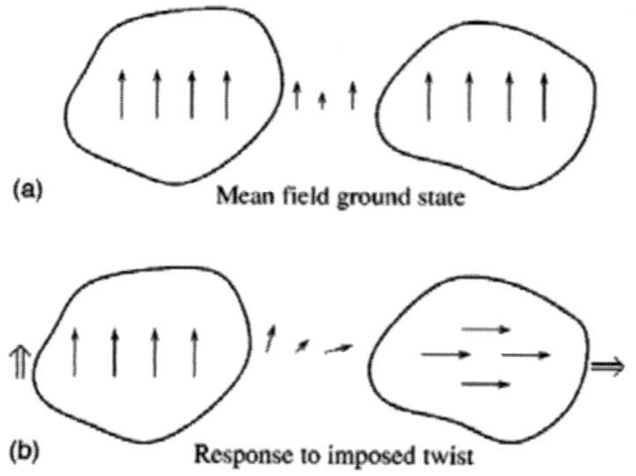

Figure 5. The superfluid density is a measure of the rigidity of the phases. This rigidity is clearly reduced by thermal fluctuations. It is also reduced by amplitude variations and by quantum phase fluctuations even at $T = 0$. The upper panel shows a phase coherent ground state albeit with variations in the amplitude– large values in the SC puddles and small values in the intervening sea. For an applied twist, the system can accommodate most of the twist in regions where the amplitude is small leading to a very small cost in energy and hence a very small superfluid density.

4. Thermal Fluctuations versus Quantum Fluctuations

Thermal phase fluctuations include Goldstone modes and vortices. These may be treated using Monte Carlo simulation of a XY or Ginzburg-Landau action for a Hubbard-Stratonovich

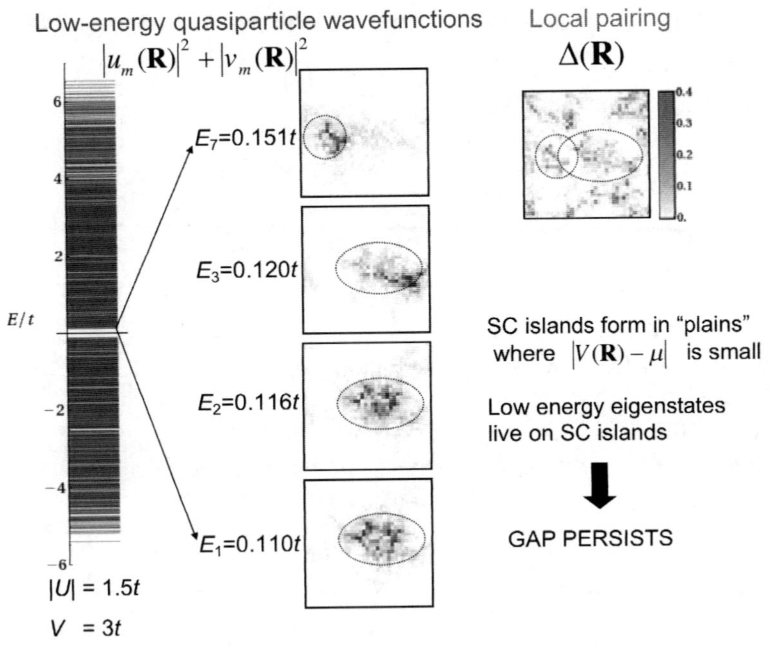

Figure 6. Eigenvalue spectrum for disorder $V = 3t$. In the absence of any attraction, the spectrum is gapless at the chemical potential. However, even the smallest attraction ($U = -1.5t$) opens up a gap. In fact, as discussed in this paper, the gap remains finite for all values of the disorder. The eigenfunctions corresponding to the eigenvalues just above the gap edge correlate extremely well with SC puddles with relatively large pairing amplitude. For a given profile of random potentials, the system chooses those regions where $|V(r) - \mu| \approx 0$ to form the SC puddles. On the other hand, eigenfunctions corresponding to the eigenvalues far below or above the gap edge correlate with deep valleys with a trapped pair or high mountains that are empty. These regions have a pairing amplitude close to zero due to strong localization effects.

auxiliary field [22, 23].

At $T = 0$ thermal fluctuations are quenched. However, long-range phase coherence can nevertheless be destroyed by quantum phase fluctuations. These can be understood to arise primarily as a consequence of the number-phase uncertainty. In a bulk superconductor, the number of fermions is completely uncertain and phase is uniform. Disorder confines the wavefunctions and places constraints on fermion number, thus introducing phase fluctuations that can ultimately drive the system into an insulator. These fluctuations can be captured qualitatively with the self-consistent harmonic approximation [20, 21, 24, 25] or more quantitatively using quantum Monte Carlo and maximum entropy methods.

5. Minimal Model for the SIT

To model the competition between superconductivity and localization that leads to the SIT in quench-condensed films with thicknesses less than the coherence length, we take the simplest lattice Hamiltonian that has an s-wave superconducting ground state in the absence of disorder ($V = 0$) and exhibits Anderson localization when the attractive interaction is turned off ($U = 0$).

Thus, we study the two-dimensional attractive Hubbard model in a random potential:

$$H = -t \sum_{\langle \mathbf{R}\mathbf{R}' \rangle \sigma} (c^\dagger_{\mathbf{R}\sigma} c_{\mathbf{R}'\sigma} + c^\dagger_{\mathbf{R}'\sigma} c_{\mathbf{R}\sigma}) - \sum_{\mathbf{R}\sigma} (\mu - V_{\mathbf{R}}) n_{\mathbf{R}\sigma} - |U| \sum_{\mathbf{R}} n_{\mathbf{R}\uparrow} n_{\mathbf{R}\downarrow}. \quad (1)$$

with lattice sites $\mathbf{R}$ and $\mathbf{R}'$, spin indices $\sigma = \uparrow$ or $\downarrow$, fermion creation and annihilation operators $c^\dagger_{\mathbf{R}\sigma}$ and $c_{\mathbf{R}\sigma}$, number operators $n_{\mathbf{R}\sigma} = c^\dagger_{\mathbf{R}\sigma} c_{\mathbf{R}\sigma}$, hopping t between neighboring sites $\langle \mathbf{R}\mathbf{R}' \rangle$, and a chemical potential μ chosen such that the average density is $\langle n \rangle \neq 1$. $V_{\mathbf{R}}$ is a random potential at each site drawn from the uniform distribution on $[-V, +V]$, and $|U|$ is the on-site attraction leading to s-wave SC. We will measure all energies in units of t.

6. Predictions from Theory

Numerous studies taken together, including Bogoliubov-de Gennes (BdG) mean-field theory, determinant quantum Monte Carlo (QMC) simulations, and maximum entropy method (MEM) analytic continuation, give rise to the following predictions near the disorder-tuned SIT.

6.1. Emergent Granularity

There has been considerable debate about whether the transition is driven by a destruction of the pairing amplitude or the phase stiffness. It was believed that homogeneously disordered films would show a closing of the single particle gap in tunneling signaling a destruction of the pairing amplitude, whereas intentionally grown granular films would lose global superconductivity because of a loss of phase coherence between grains even though the individual grains could be locally superconducting. Our results indicate that, at least for this minimal model, there is a *single paradigm* for homogeneously disordered and granular films. For low disorder the local pairing amplitude $\Delta(\mathbf{R}) \equiv \langle c_{\mathbf{R}\downarrow} c_{\mathbf{R}\uparrow} \rangle$ is homogeneous across the system. However, as disorder is increased the system self-organizes into superconducting blobs on the scale of the coherence length within an insulating matrix (see Figs. 4 and 6). The phases of the different blobs are coupled by Josephson tunneling of pairs. In the globally superconducting state, the phases of the different blobs get locked together whereas in the insulator the phase coherence of the different blobs is lost on ever shorter length and time scales as one moves away from the quantum phase transition.

6.2. Pairing amplitude does not equal gap

In a clean system, the pairing amplitude equals the single-particle gap, and it is common to use the symbol Δ for both quantities.

In a disordered system near the SIT, the pairing amplitude $\Delta(\mathbf{R})$ is strongly inhomogeneous. Does it describe the local gap $E_{\text{gap}}(\mathbf{R})$ as measured by scanning tunneling spectroscopy? The answer is no!

Figure 7 shows that there is in fact an anti-correlation between the local pairing amplitude and the local spectral gap. The SC puddles on which the local $\Delta(\mathbf{R})$ is finite have a finite gap with symmetric line shapes and sharp coherence peaks or pile-ups in the density of states at the gap edges. On the other hand, the insulating regions have $\Delta(\mathbf{R}) \approx 0$ and very asymmetric broad density of states showing a much larger gap. Although the local gap extracted from the local density of states (LDOS) is highly inhomogeneous, it is nevertheless finite at every site, similar to the experimental data in Fig. 8.

The DOS is the LDOS averaged over all sites. The gap in the DOS, E_{gap}, is the lowest gap in the LDOS on any site. According to BdG (Fig. 10) and QMC (Fig. 14) calculations, E_{gap}

Journal of Physics: Conference Series **376** (2012) 012001 doi:10.1088/1742-6596/376/1/012001

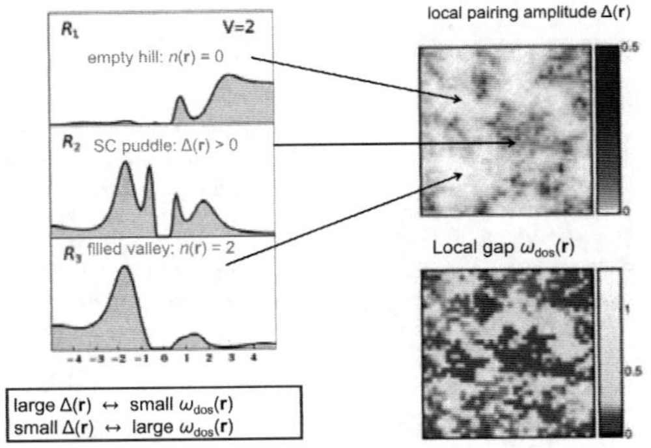

Figure 7. (Right) Within BdG, the local pairing amplitude is *anticorrelated* with the local gap. (Left) LDOS results from QMC+MEM. Site $\mathbf{R_1}$ is on a high potential hill that is nearly empty, and $\mathbf{R_3}$ is in a deep valley that is almost doubly occupied. This leads to the characteristic asymmetries in the LDOS for $\mathbf{R_1}$ and $\mathbf{R_3}$. The small local pairing amplitude $\Delta(\mathbf{R})$ at these two sites is reflected in the absence of coherence peaks in the LDOS. In contrast, site $\mathbf{R_2}$ has a density closer to half-filling, leading to a significant local pairing amplitude, a much more symmetrical LDOS, and coherence peaks that persist even at strong disorder.

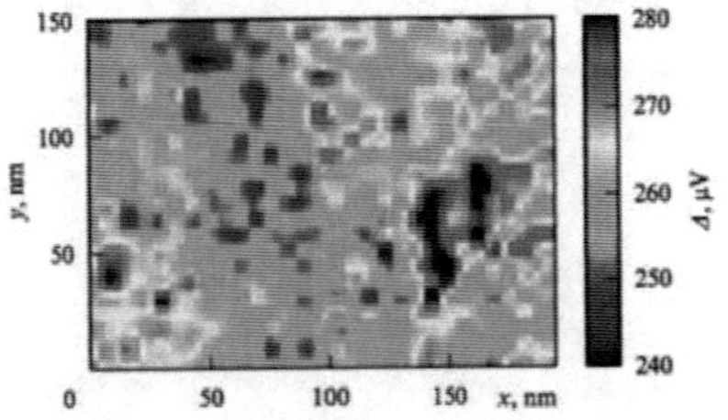

Figure 8. Gap map of a TiN film obtained from scanning tunneling spectroscopy, showing inhomogeneities on a scale of a few tens of nanometers.

remains robust across the SIT, even when thermal and quantum phase fluctuations are included. Thus the SIT is a transition from a gapped superconductor to a gapped insulator.

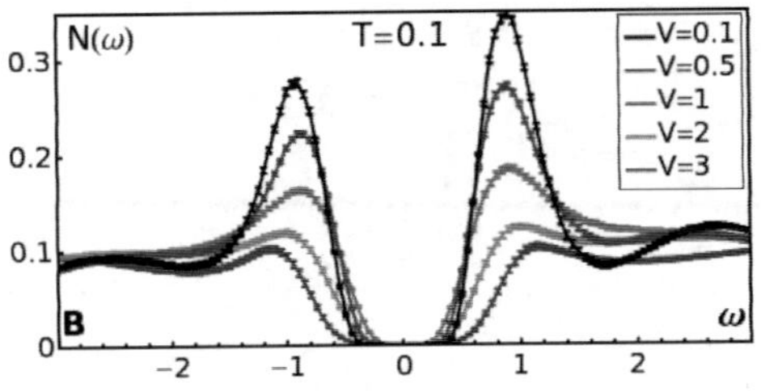

Figure 9. Disorder dependence of single-particle spectrum from QMC+MEM at very low temperature [26]. With increasing disorder, quantum phase fluctuations eventually wash out the coherence peaks, but the gap is robust.

6.3. Coherence peaks disappear beyond SIT

These characteristic pile-ups in the DOS at the gap edges appear to be directly correlated with superconducting order. They vanish as the disorder is increased across the SIT (Fig. 9), or as the temperature is raised above T_c (Fig. 11). These predictions agree very well with experiments (see Fig. 12).

6.4. Pseudogap over wide temperature range

Near the SIT, a pseudogap – a suppression in the low-energy DOS – persists well above the superconducting T_c up to a crossover temperature scale T^*, in marked deviation from BCS

Journal of Physics: Conference Series **376** (2012) 012001 doi:10.1088/1742-6596/376/1/012001

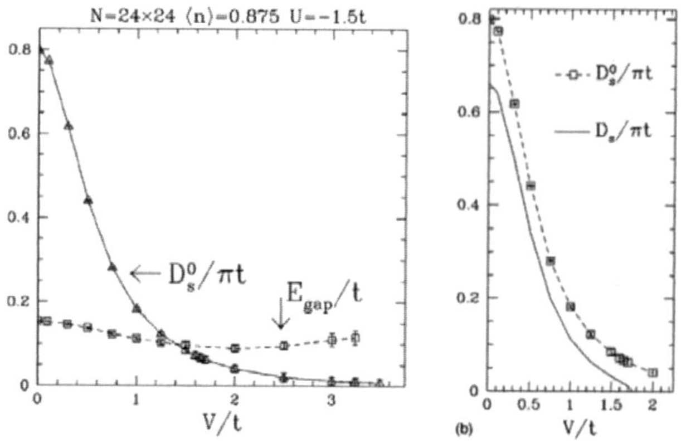

Figure 10. (Left) According to BdG simulations, the gap persists across the SIT. This conclusion holds even after phase fluctuations are taken into account. (Right) The renormalized superfluid stiffness D_s, including phase fluctuations within the SCHA, falls to zero at the SIT.

theory. This disorder-driven pseudogap also exists at finite temperatures in the insulating state and grows with disorder (Fig. 11.) These predictions are again in good agreement with experiments (Fig. 12).

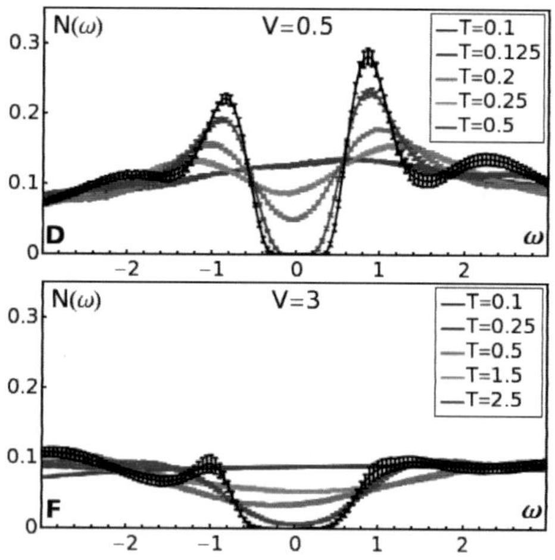

Figure 11. Temperature dependence of DOS from QMC+MEM calculations [26]. (Top) At weak disorder, as a function of increasing temperature, thermal fluctuations destroy the coherence peaks for $T \gtrsim T_c \approx 0.14$. However, a pseudogap remains up to higher temperatures $T \sim 0.4$. (Bottom) At strong disorder, there are no coherence peaks; there is a hard gap at $T = 0$ and a pseudogap up to $T \sim 1.5$.

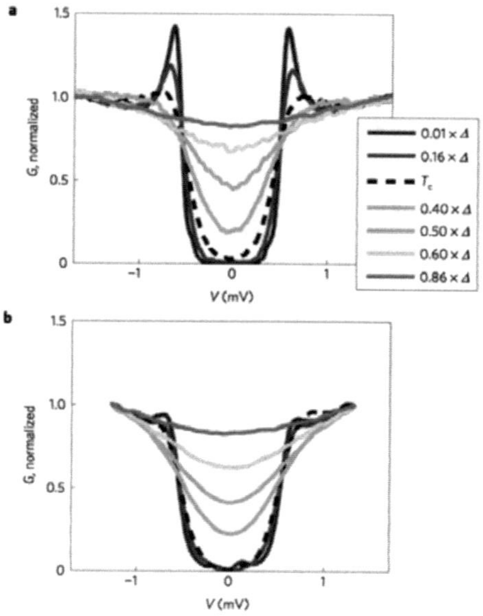

Figure 12. Local tunneling conductance spectrum in InO films [27]. (Top) At low disorder there are coherence peaks below T_c and a pseudogap up to much higher T. (Bottom) At high disorder there are no coherence peaks, but nevertheless there is a hard gap at low T and a pseudogap up to high temperatures.

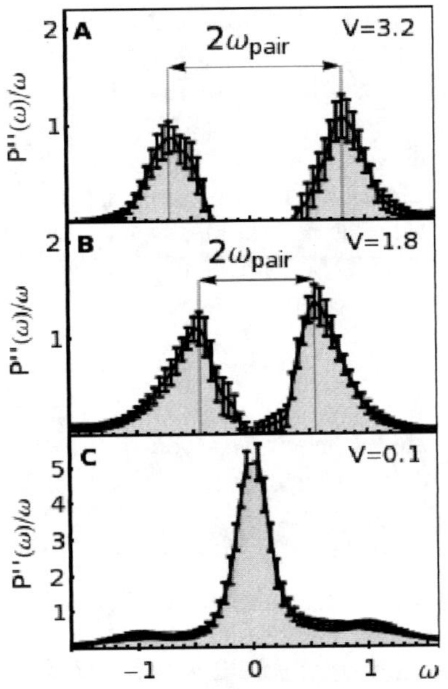

Figure 13. Imaginary part of the dynamical pair susceptibility $P''(\omega)/\omega$ at $T = 0.1t$. Error bars represent variations between 10 disorder realizations. For $V < V_c$ the large peak at $\omega = 0$ indicates zero energy cost to insert a pair into the SC. For $V > V_c$, there is a gap-like structure at $\pm\omega_{\text{pair}}$, the typical energy required to insert a pair into the insulator.

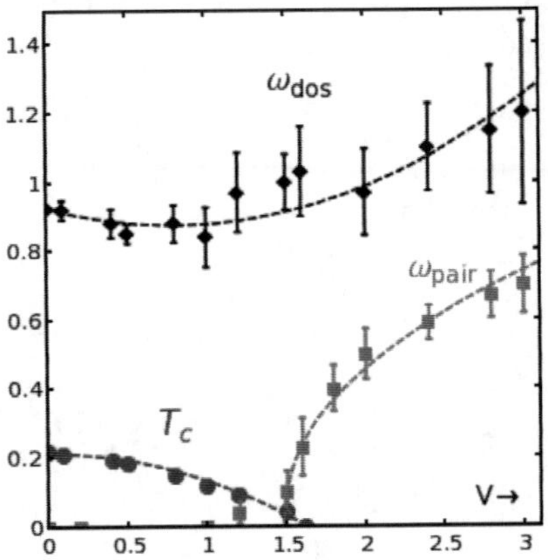

Figure 14. Energy and temperature scales across the superconductor-insulator transition (SIT) according to QMC+MEM calculations. The single-particle gap ω_{dos} remains finite for all values of disorder V, whereas the superconducting T_c and the two-particle energy scale ω_{pair} in the insulator both vanish at the SIT.

6.5. Two-particle spectrum

The local two-particle spectral function, or pair susceptibility $P(\omega)$, is defined as the analytic continuation of the correlation function $P(\tau) = \sum_{\mathbf{R}} \langle T_\tau F(\mathbf{R};\tau) F^\dagger(\mathbf{R};0)\rangle$ where $F(\mathbf{R},\tau) = c_{\mathbf{R}\downarrow}(\tau) c_{\mathbf{R}\uparrow}(\tau)$. Physically, $P(\mathbf{R},\omega)$ is the amplitude for inserting a pair at a site $\mathbf{R}$ at energy ω, and $P(\omega)$ is the average insertion amplitude over all sites.

Fig. 13 shows QMC+MEM results for the imaginary part of P. On the superconducting side of the transition there is a large amplitude for inserting pairs at zero energy. However, on the insulating side, there is a characteristic energy scale ω_{pair} to insert a pair in the insulator that collapses upon approaching the SIT (notwithstanding a small amount of spectral weight at low energies coming from rare regions).

6.6. Suggestions for Further Experiments

A large number of experiments exploring SITs in conventional s-wave materials have focused on transport measurements. In order to get deeper insights into the phases, it is useful to bring to bear a larger variety of experiments of SIT, such as a detailed and systematic analysis of scanning tunneling spectroscopy and dynamical conductivity, some of which have already begun. In the cuprates, considerable insight about the pseudogap region was obtained by measuring the entropy carried by vortices from the Nernst effect [28]. Similar experiments for the regular SIT should also be extremely useful [29]. It is evident that there should be large diamagnetic effects in the insulator because of puddles of superconducting regions embedded in an insulating matrix. Once again direct probes of diamagnetic in the insulator and its evolution toward the SIT, similar to the explorations in the cuprates[30, 31], should be extremely instructive. Unlike the disorder driven transition, the effect of a perpendicular magnetic field is not as well understood. The experiments show that the sharp normal to SC transition at T_c for zero field develops long tails in the presence of a magnetic field. In other words there appears to be some source of dissipation in the SC state. Is this due to unpinned vortices? Why is the disorder, even when it is as high as the quantum of resistance, unable to pin the vortices?

And finally, once the phases are understood, the next frontier will be investigations of the quantum phase transition. In this context the ability to tune across the transition by gating is a breakthrough! Gating allows one to separately tune the effects of screening and disorder and will hopefully provide a much better understanding of quantum criticality.

6.7. Acknowledgments

We gratefully acknowledge support from US Department of Energy, Office of Basic Energy Sciences grant DOE DE-FG02-07ER46423 (N.T., Y.L.L.), NSF DMR-0907275 (K.B.), NSF DMR-0706203 and NSF DMR-1006532 (M.R.), and computational support from the Ohio Supercomputing Center.

References

[1] Anderson P W 1959 *Journal of Physics and Chemistry of Solids* **11** 26–30
[2] Ma M and Lee P A 1985 *Phys. Rev. B* **32** 5658–5667
[3] Strongin M, Thompson R S, Kammerer O F and Crow J E 1970 *Phys. Rev. B* **1** 1078–1091
[4] Haviland D B, Liu Y and Goldman A M 1989 *PRL* **62** 2180
[5] Valles J M, Dynes R C and Garno J P 1989 *PRB* **40** 6680
[6] Valles J M, Dynes R C and Garno J P 1992 *PRL* **69** 3567
[7] Hebard A F and Paalanen M A 1990 *Phys. Rev. Lett.* **65** 927–930
[8] Shahar D and Ovadyahu Z 1992 *Phys. Rev. B* **46** 10917–10922
[9] Chervenak J A and Valles J M 1999 *PRB* **59** 11209
[10] Steiner M A, Boebinger G and Kapitulnik A 2005 *Phys. Rev. Lett.* **94** 107008
[11] Stewart M D, Yin A, Xu J M and Valles J M 2007 *Science* **318** 1273
[12] Nguyen H Q, Hollen S M, Stewart M D, Shainline J, Yin A, Xu J M and Valles J M 2009 *PRL* **103** 157001
[13] Lee Y, Clement C, Hellerstedt J, Kinney J, Kinnischtzke L, Leng X, Snyder S D and Goldman A M 2011 *Phys. Rev. Lett.* **106** 136809
[14] Lin Y H and Goldman A M 2011 *Phys. Rev. Lett.* **106** 127003
[15] Bollinger A T, Dubuis G, Yoon J, Pavuna D, Misewich J and Bozovic I 2011 *Nature* **472** 458–460 ISSN 0028-0836
[16] Randeria M, Trivedi N, Moreo A and Scalettar R T 1992 *Phys. Rev. Lett.* **69**(13) 2001–2004
[17] Trivedi N and Randeria M 1995 *Phys. Rev. Lett.* **75**(2) 312–315
[18] Randeria M *The Crossover from BCS theory to Bose-Einstein Condensation*, in *Bose-Einstein Condensation*, ed. A. Griffin, D. Snoke, and S. Stringari, p. 355–392, Cambridge University Press (1995).
[19] Randeria M 2010 *Nat. Phys.* **6** 561–562 ISSN 1745-2473
[20] Ghosal A, Randeria M and Trivedi N 1998 *PRL* **81** 3940
[21] Ghosal A, Randeria M and Trivedi N 2001 *Phys. Rev. B* **65** 014501
[22] Dubi Y, Meir Y and Avishai Y 2007 *Nature* **449** 876–880 ISSN 0028-0836
[23] Dubi Y, Meir Y and Avishai Y 2008 *PRB* **78** 024502
[24] Ramakrishnan T V 1989 *Phys. Scr.* **T27** 24
[25] Wood D and Stroud D 1982 *Phys. Rev. B* **25** 1600
[26] Bouadim K, Loh Y L, Randeria M and Trivedi N 2011 *Nature Physics* **7** ISSN 884–889
[27] Sacépé B, Dubouchet T, Chapelier C, Sanquer M, Ovadia M, Shahar D, Feigel'man M and Ioffe L 2011 *Nat. Phys.* **7** 239–244 ISSN 1745-2473
[28] Wang Y, Ono S, Onose Y, Gu G, Ando Y, Tokura Y, Uchida S and Ong N P 2003 *Science* **299** 86–89 (*Preprint* `http://www.sciencemag.org/content/299/5603/86.full.pdf`)

[29] Pourret A, Aubin H, Lesueur J, Marrache-Kikuchi C A, Berge L, Dumoulin L and Behnia K 2006 *Nat. Phys.* **2** 683–686 ISSN 1745-2473
[30] Wang Y, Li L, Naughton M J, Gu G D, Uchida S and Ong N P 2005 *Phys. Rev. Lett.* **95** 247002
[31] Li L, Wang Y, Komiya S, Ono S, Ando Y, Gu G D and Ong N P 2010 *Phys. Rev. B* **81** 054510

14th International Conference on Transport in Interacting Disordered Systems (TIDS-14) IOP Publishing
Journal of Physics: Conference Series **376** (2012) 012002 doi:10.1088/1742-6596/376/1/012002

Cooper pair islanding model of insulating nanohoneycomb films

S. M. Hollen

J. M. Valles, Jr.

Department of Physics, Brown University, Providence, RI 02912

Abstract.

We first review evidence for the Cooper pair insulator (CPI) phase in amorphous nanohoneycomb (NHC) films. We then extend our analysis of superconducting islands induced by film thickness variations in NHC films to examine the evolution of island sizes through the magnetic field-driven SIT. Finally, using the islanding picture, we present a plausible model for the appearance and behavior of the CPI phase in amorphous NHC films.

Journal of Physics: Conference Series **376** (2012) 012002 doi:10.1088/1742-6596/376/1/012002

1. Introduction

The appearance of a giant magnetoresistance peak near the superconductor to insulator transition (SIT) in several thin film systems[1, 2, 3, 4, 5] has garnered a lot of recent attention.[6, 3, 7] The factors that drive its appearance are not yet understood. Of the thin film systems that exhibit this exotic behavior, nanohoneycomb (NHC) amorphous Bi thin films offer a unique morphology that might lead to an explanation.

The amorphous NHC thin films considered here were fabricated by quench condensing Sb and then Bi onto a substrate of Anodized Aluminum Oxide (AAO) (see Fig. 2a). In the zero-temperature limit, an insulator-superconductor transition can be driven by increasing the NHC film thickness, or a superconductor-insulator transition can be driven by applying a magnetic field to a superconducting film. Both of these types of transitions bear little resemblance to their counterparts in films grown on planar glass substrates, even though we can be sure that films grown on AAO are locally homogeneous.[8] Most interestingly, the insulating phase in NHC films contains localized Cooper pairs, as confirmed by the observation of Little-Parks like oscillations in the magnetoresistance.[9] Recently, the localization of Cooper pairs has been attributed to the formation of superconducting islands near the SIT that develop due to nano-scale undulations in the film thickness.[8] The appearance of these localized Cooper pairs is likely responsible for the puzzling features in the transport, including the giant magnetoresistance peak appearing at fields beyond the Little-Parks oscillations (1-3T).[10]

In this paper we examine the transport behavior of amorphous NHC films near the SIT within the context of this Cooper pair islanding model. We present images indicating how increasing film thickness or magnetic field changes the island size. Inspired by the somewhat regular arrangement of the islands in the images, we use an ordered grain array model to qualitatively explain how the evolution of island sizes can lead to the spectacular transport behavior in the Cooper pair insulator phase of these films.

2. Amorphous NHC film transport near the SIT

NHC thin films on the insulating side of the SIT show simply activated behavior, where $R_{\square}(T) = R_0 exp(T_0/T)$, whether they are driven by disorder or an applied magnetic field. Additionally, the activation energy, T_0, goes continuously to zero at the transition in both cases. Fig 1a shows a typical thickness-driven insulator to superconductor transition for NHC films on Arrhenius plot. The linearly decreasing activation energies of film sequences from four different experiments are shown in Fig 1b (triangles are T_0 data extracted from the series in Fig 1a). In this figure, the transition to superconductivity is marked by $d^{\mathrm{dep}}(T_0 = 0)$.

Complementary data for the field-driven superconductor-insulator transition are shown in Figs. 2c and d. The activation energy (Fig. 2d) changes dramatically in

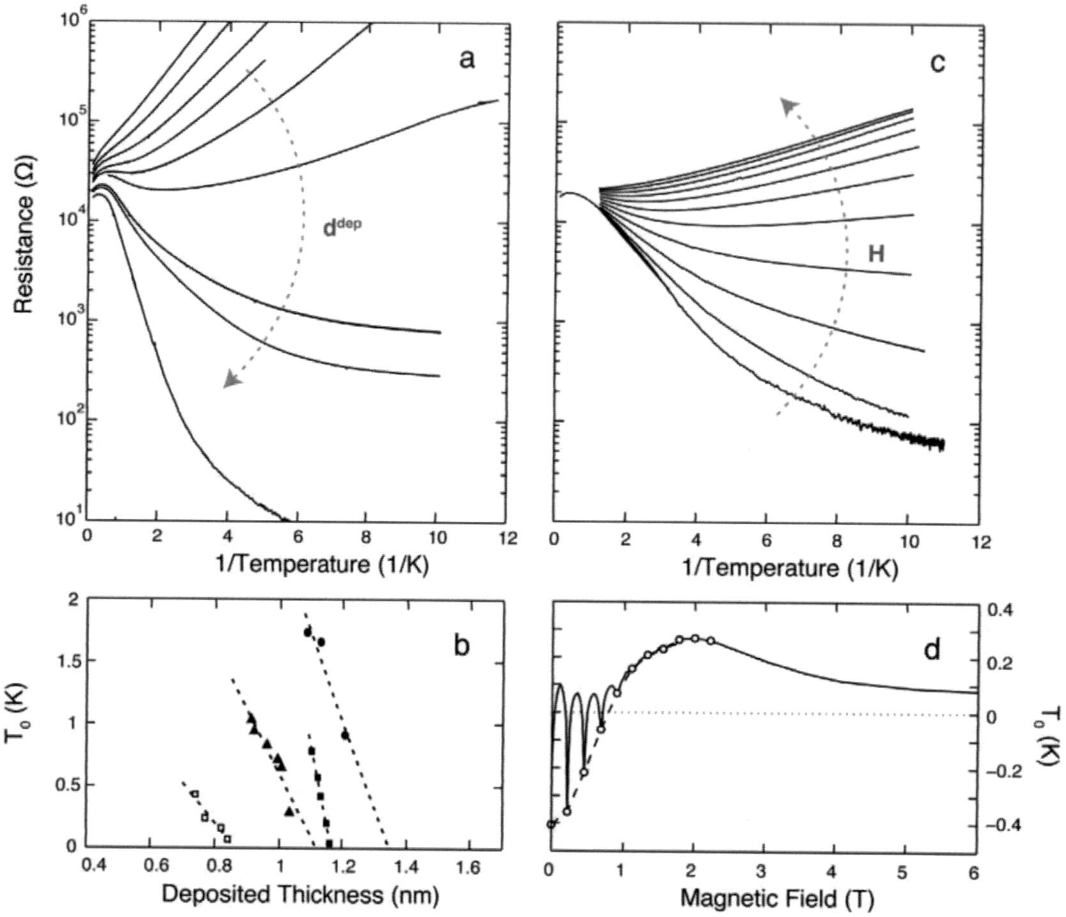

Figure 1. a) Thickness-driven IST in NHC films on an Arrhenius plot. From top to bottom, the deposited film thicknesses are 0.91, 0.92, 0.96, 0.99, 1.01, 1.03, 1.05, 1.08, 1.12 nm. b) Activation energies, obtained from the slopes of Arrhenius fits, versus deposited thickness for four different series of NHC films. Triangles represent T_0 extracted from data in (a). c) Perpendicular field-driven SIT of a 1.13 nm NHC film on an Arrhenius plot. Field values from bottom to top are 0, 0.22, 0.45, 0.68, 0.90, 1.11, 1.33, 1.55, 1.78, 2.00, 2.22 T. d) Activation energy versus applied field for the film in (c). The $R(T)$ of the open circles (at $H = nH_M$, $H_M = 0.22$ T) are shown in (c). Negative energy values are plotted for consistency; a dashed line divides negative and positive slopes (superconducting and insulating states).

magnetic field, mimicking the behavior of the magnetoresistance with both oscillations and a peak in T_0.[10, 9] In contrast, the prefactor, R_0, increases monotonically by a factor of 2 through the entire field range (not shown).[10] Because of the oscillations, the initially superconducting film undergoes 7 consecutive transitions between superconducting and insulating states (marked by crossing the $T_0 = 0$ line). The $R(T)$ curves transversing the SIT in c are for the field values chosen at $H = nH_{\rm M}$ to avoid the intra-oscillation transitions (open circles in d). The form of $T_0(H)$ suggests two separate contributions to the activation energy:

$$T_0(H) = T_0^{\rm osc}(H) + T_0^{\rm peak}(H). \qquad (1)$$

The origin of the overarching $T_0^{\rm peak}$ feature is not yet determined, but a similarly shaped feature appears in the other highly-disordered (unpatterned) film systems, mentioned in the introduction and in more detail in the discussion.

At low fields, we expect that $T_0^{\rm osc}$ is dominated by a Josephson energy between islands, which oscillates with the matching field period, as in Josephson Junction Arrays (JJAs).[11] The following expression is often used to characterize this oscillating energy scale,

$$T_0^{\rm osc}(H) \propto < E_{J0} cos(\phi_i - \phi_j - 2\pi A_{ij}/\Phi_0) >, \qquad (2)$$

where E_{J0} is the zero-field phase coupling energy, the ϕs are phases on neighboring islands in the film, A_{ij} is line integral of the field's vector potential between islands, and $\Phi_0 = h/(2e)$ is the magnetic flux quantum. The presence of this term in the activation energy is strong evidence that activated Cooper pair tunneling dominates the transport.

3. Cooper pair islanding in NHC films

A comparison of NHC and uniform films with increasing deposition thickness intimates that structure in the AAO substrates leads to Cooper pair localization. The AAO is not totally flat: regular height variations appear around the holes, as can be seen in the atomic force microscope image in Fig. 2a. Since evaporated material impinges normal to the (average) plane of the substrate, thickness variations develop according to the local slope:

$$d(x,y) = d^{\rm dep}\frac{1}{\sqrt{1+(\nabla h(x,y))^2}}, \qquad (3)$$

where $d(x,y)$ is the local film thickness and $h(x,y)$ is the local height of the substrate. Evidently, the film grows thickest in the flat regions of the substrate and thinnest in the steep regions. Fig. 2b shows the height profile along a line scan of the substrate as well as the corresponding thickness profile, calculated using Eqn. 1 and $d^{\rm dep} = 0.89$nm (an insulating film). Also on this plot, we mark the critical thicknesses for conduction, $d_{\rm Gc}^{\rm ref}$, and superconduction, $d_{\rm IST}^{\rm ref}$, in reference films. It is evident that some regions of the film are thick enough to support superconductivity while others are not.

By applying Eqn. 1 to the AFM image of Fig 2a, we create a map of the local film thickness for a given $d^{\rm dep}$. In addition, we color the map so that regions of film that conduct ($d_{\rm Gc}^{\rm ref} < d(x,y) < d_{\rm IST}^{\rm ref}$) are pink, and regions that superconduct ($d(x,y) > d_{\rm IST}^{\rm ref}$) are blue. Non-conducting film ($d(x,y) < d_{\rm Gc}^{\rm ref}$) is gray. Fig. 2c shows the resulting morphology of an insulating film near the IST ($d^{\rm dep} = 0.89$nm). Here, regions of superconductivity are interspersed in regions of simply conducting film.

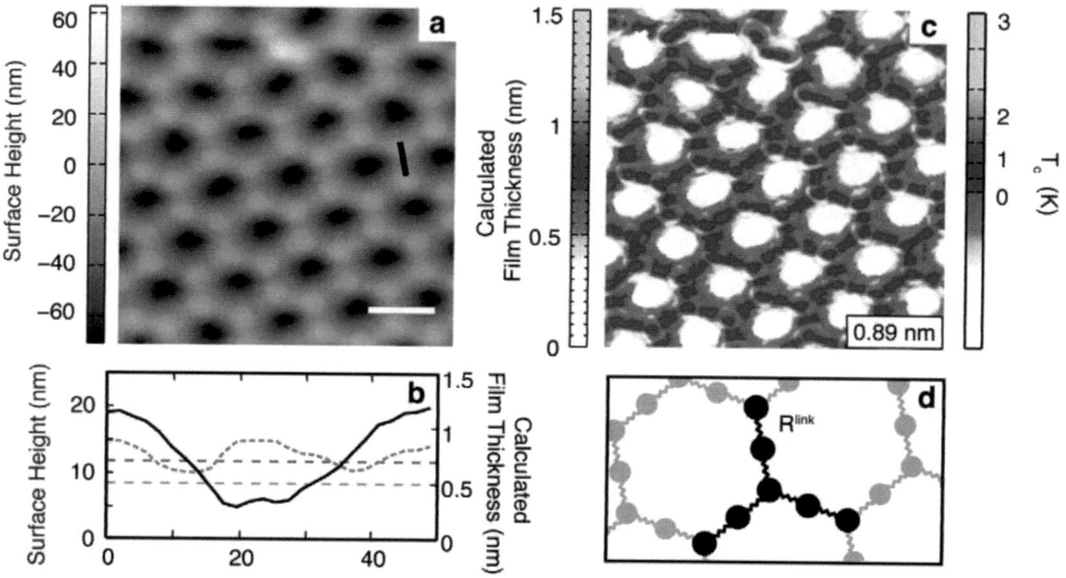

Figure 2. Figure adapted from reference [8] a) AFM image of the AAO substrate. The (white) scale bar spans 100 nm. b) Height profile of substrate along a line scan (black line in (a)). The dashed gray curve is the corresponding film thickness profile calculated using Eqn. 3. Pink and blue dashed lines mark the critical thicknesses for conductivity and superconductivity, respectively. c) Film thickness variations calculated using the image in (a) and Eqn. 3. Colored according to critical thicknesses in a reference film: pink if the film is thick enough to conduct ($d^{\rm dep} > d^{\rm ref}_{\rm Gc} = 0.5$nm), and blue if the film is thick enough to superconduct ($d^{\rm dep} > d^{\rm ref}_{\rm IST} = 0.7$nm). The right color bar defines the local coupling constant, $T_{\rm c}$, calculated using Eqn. 4. d) A model of the film as superconducting islands connected by resistor elements, or weak links.

Taking the analysis one step further, we can transform the film thickness variation map into a coupling constant map, again by comparing to data of reference films. In uniform superconducting films, the transition temperature depends on film thickness as,

$$T_{\rm c}^{\rm ref} = T_{\rm c}^{\rm bulk}(1 - d^{\rm ref}_{\rm IST}/d^{\rm dep}). \tag{4}$$

For a-Bi, $T_{\rm c}^{\rm bulk} = 6$ K and $d^{\rm ref}_{\rm IST} = 0.7$nm.(Jay thesis) Assuming this relation to be true on a microscopic level, we add a color bar to the right side of Fig. 2b to convert our map of local film thicknesses to one of local $T_c s$. Considering the picture as a whole, the film appears to be made up of an array of superconducting islands connected by non-superconducting film. The local coupling constant, marked by a local $T_{\rm c}$, varies over an island. It is strongest in the center, where $d(x,y)$ is largest, and weakest on the edges. By analyzing these maps through the IST, we find that the superconducting islands grow with $d^{\rm dep}$ and coalesce at the transition.[8]

Viewing the transitions as the coalescence of conducting or superconducting regions of film allows us to understand the critical deposition thicknesses for conduction ($d^{\rm NHC}_{\rm Gc}$) and superconduction ($d^{\rm NHC}_{\rm SC}$) in NHC films.[8] This result is corroborated by modeling

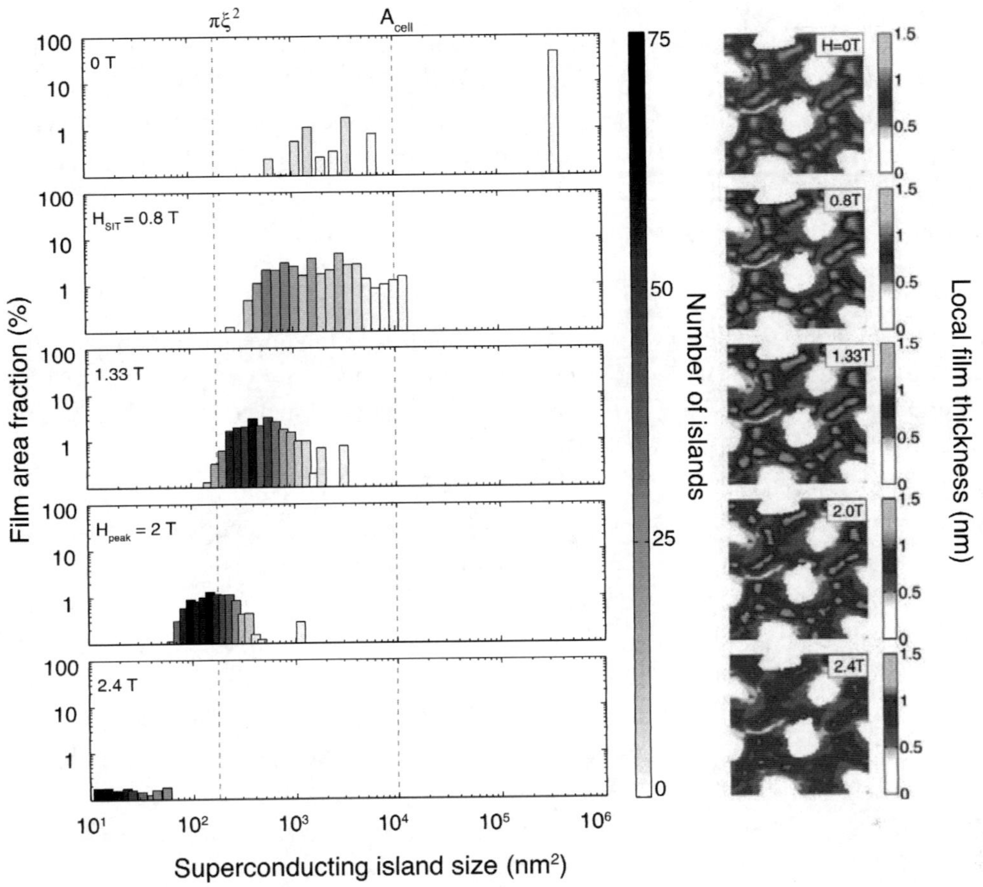

Figure 3. Superconducting islands in an increasing magnetic field for the 1.08nm film of Fig 1c,d. Right: Spatial map of SC islands (blue) in an insulating background (pink) with increasing magnetic field. The cranberry color represents regions of initially superconducting film for which $H_{c2}(d(x,y)) \leq H_{\text{applied}}$, calculated using Eqn. 5. Left: Distribution of superconducting island size and number of islands of each size for the applied field series. The first, third, and fourth panels (from the top) correspond to R(T) shown in Fig. 1c.

the film as a network of islands linked by resistor elements, shown in Fig 2d. At the SIT, the resistance of a single weak link in this network is consistently close to $R_Q = h/4e^2$, the critical value of sheet resistance in uniform films. Both of these observations indicate that the weak links between superconducting islands control the transition to superconductivity in these films.[8]

4. Shrinking Cooper pair islands with an applied magnetic field

We can extend our islanding picture to undulating films in an applied magnetic field by once again comparing the local film thicknesses to uniform films. This fairly primitive approach has led to a nice picture that qualitatively agrees with some models described in the discussion. We attribute the critical field in uniform amorphous Bi films of various thicknesses, $H_{c2}(d^{\text{dep}})$, to local film thicknesses in undulating amorphous Bi films. In uniform films, H_{c2} depends on thickness as,

$$H_{c2} = 7.12\text{T}(1 - 0.7\text{nm}/d). \tag{5}$$

(determined by a fit to the data)[12]. Using this phase line, we can define a critical thickness for superconductivity for every value of applied field, $d_{SIT}(H)$. In Fig. 4, the superconducting film of Fig. 1c,d ($d^{\text{dep}} = 1.08$nm) is shown for a series of applied magnetic fields. At each field value the color bar is redefined to include a region of previously superconducting film made normal by the applied field (cranberry colored). In the figure, the islands shrink with increasing field, eventually disappearing at H=2.5T (not shown). The accompanying histograms show the magnetic field evolution of the island size distribution, as the number of islands of each size (in a 1x1μm region), and the percent film area taken up by islands of that size. At H=0T, a single island dominates the film area, but as the field is increased these islands separate. The second panel from the top shows the rather broad distribution of islands at the field-driven SIT (0.8T), where they have just separated. At H_{peak}=2T, the islands are much smaller and fairly normally distributed in number and area around the estimated coherence area, $\pi\xi^2$. Beyond the peak, at H=2.4T, the islands are all smaller than $\pi\xi^2$ and have nearly disappeared.

5. Discussion

Dramatic transport behavior near the SIT, similar to what has been presented here (see Fig. 1), has been observed in a number of thin film systems, including indium oxide, titanium nitride, and beryllium.[1, 4, 13] In particular, a number of models have been formulated to try to explain the origin of the giant magnetoresistance peak.[6, 14, 7, 3] In these models, the peak is commonly attributed to the existence of islands of incoherent Cooper pairs. Our analysis of the morphology of NHC films suggests a well defined mechanism for island formation. This mechanism provides a picture of the geometrical arrangement and rough sizes of the islands through the disorder and magnetic field tuned SITs. Below, we combine these unique insights with and ordered grain array model to provide a qualitative explanation for the activated insulator and MR peak.

5.1. The hard gap at the SIT

A number of models produce energy scales related to Cooper pairs in the insulator phase that may correspond to the activation energy seen in insulating NHC films (see

Fig. 1b,d). Trivedi and coworkers[15] showed that an attractive Hubbard model with disorder predicts a pair gap that opens at the SIT and grows into the insulating state. This gap corresponds to the energy required to inject a Cooper pair into the insulator. Very recently, Müller[14, 16] proposed that a mobility edge can appear in a disordered, localized bosonic system with Coulomb interactions. The distance to the mobility edge, which serves as the activation energy for Cooper pair transport, increases with disorder and magnetic field. Finally, and of primary interest here, an ordered array of small, Josephson coupled, superconducting grains is expected to exhibit a Mott insulator phase with an energy barrier for Cooper pair injection (see [17] and refs. therein). In the limit of very weak coupling between grains this barrier corresponds to the self charging energy for a grain, E_{C}. Close to the SIT, the increased coupling between grains, as measured by the Josephson energy, E_{J}, leads to a downward renormalization of this barrier, $E_{\mathrm{C}} \rightarrow \tilde{E_{\mathrm{C}}}$. Deep in the insulator, $\tilde{E_{\mathrm{C}}} \simeq E_{\mathrm{C}} - E_{\mathrm{J}}/4$. As the intergrain normal state resistance approaches R_{Q}, this renormalization can be very substantial.[17]

The network of islands that results from CPs localized by coupling constant variations suggests applying the ordered grain array model to NHC films. Within this model, we estimate that the islands contain few Cooper pairs and have a large energy level spacing. The bare charging energy between grains, E_{C}, is found to be enormous for the islands in Fig. 2 (approximately 100 meV)[9] and much greater than the Josephson energy, E_{J}, so that the applicability of the model would seem to rely on a very strong renormalization of E_{C} near the SIT.

The number of Cooper pairs in a coherence volume can be estimated as $N_{\mathrm{CP}} = V_{\mathrm{island}}\nu(\epsilon_F)\Delta$, where V_{island} is the volume of the superconducting island, $\nu(\epsilon_F)$ is the density of states at the Fermi energy and Δ is the amplitude of the superconducting order parameter. Taking an island near the SIT to be 20nm in diameter and 1nm thick, $V_{\mathrm{island}} \approx 100\pi\mathrm{nm}^3$. We estimate $\nu(\epsilon_F)$ for a-Bi to be $2 \times 10^{22}\mathrm{cm}^{-3}\mathrm{eV}^{-1}$, the value for crystalline Pb. Finally, we estimate Δ using $2\Delta = 3.5k_BT_c$ and the local maximum T_c from the coupling constant map (Fig. 2c). Using these numbers we find that $N_{CP} \approx 2$. Additionally, the energy level spacing in an island, $\delta = 1/\nu(\epsilon_F)V_{\mathrm{island}}$, yields $\delta \approx 1.76\mathrm{K}$.

To estimate the Josephson energy, we use $E_{\mathrm{J}} = \pi g\Delta/2$ where the normalized conductance, $g \approx 1$ near the transition. For $T_{\mathrm{c}} = 2\mathrm{K}$, this expression gives $E_{\mathrm{J}} \approx 5\mathrm{K}(\approx 0.4\mathrm{meV})$. Finally, we calculate the charging energy with $E_{\mathrm{C}} = 4e^2/\epsilon_0\epsilon L$ using $L = 20\mathrm{nm}$, as the superconducting island size (see Fig. 2c). We find $E_{\mathrm{C}} \approx 10^4\mathrm{K}/\epsilon$, implying $E_{\mathrm{C}} \gg \delta, T_{\mathrm{c}}, E_{\mathrm{J}}$ for any reasonable ϵ.

Near the SIT, the condensation energy per grain, the Josephson energy, the energy level spacing, and T_{c} are all of the same order. The charging energy per grain is very large, however. We expect it to be dominant in this model even after the necessary renormalization. On approaching the disorder-driven SIT from the insulating side, the superconducting islands grow with d^{dep} and coalesce at the transition.[8] As the islands grow, their separation shrinks and they become better coupled. These changes result in a diminishing E_{C} and an increasing E_{J} as $d^{\mathrm{dep}} \rightarrow d_{IST}^{\mathrm{NHC}}$. The growing islands accommodate increasingly more Cooper pairs. These changes result in a reduction of

$\tilde{E_C}$, which corresponds to the decreasing $T_0(d^{dep})$ in Fig. 1b.

5.2. *The magnetoresistance peak*

Here, we extend the discussion of NHC films modeled as an ordered island array to account for the effects of an applied magnetic field on their transport. With Eqn. 2, we attribute the influence of the magnetic field on the activation energy to a combination of two terms. The low field magneto-oscillations in T_0^{osc} (Fig. 1d) result from field-induced modulations of E_J and, as a result, $\tilde{E_C}$. At higher fields, T_0^{peak} dominates and gives rise to the magnetoresistance peak. T_0^{peak} grows due to the magnetic field induced shrinking of the islands, which drives up Ec and reduces the inter-island coupling, reducing E_J. Also, the magnetic field reduces Δ, which also reduces E_J. Altogether, these effects drive up $\tilde{E_C}$, or T_0^{peak}. Eventually, the magnetic field depresses Δ below $\tilde{E_C}$, at which point quasiparticle tunneling rather than CP tunneling becomes energetically favorable. As pointed out by Gantmakher [3] and others, the magnetoresistance becomes negative for quasiparticle tunneling because the tunneling rate increases as the field depresses the gap.

5.3. *Conclusion*

We reviewed evidence for the existence of a Cooper pair insulator phase in amorphous NHC films. AFM images reveal that structure in the substrates induce the formation of superconducting islands separated by weak links. We presented the evolution of these island sizes with film thickness and magnetic field. We then discussed an ordered grain array model that can qualitatively account for the thickness and magnetic field tuned SITs as well as the large magnetoresistance peak observed in the Cooper pair insulator state.

While the ordered island array scenario is plausible, experimental observations still allow other interpretations. The grain array model attributes the activation energy to charging effects, which other models do not need to include (see [18]). It also requires the array to be ordered, as disorder in the array would likely lead to variable range hopping or percolation phenomena. The model of Dubi and Meir[6] invokes magnetic field induced shrinkage of superconducting islands, but also uses a percolation conduction model because many of the systems exhibiting the MR peak are disordered arrays.

We hope our results provide a starting point for explaining the common features in the transport across seemingly dissimilar systems exhibiting Cooper pair insulating behavior. Future theoretical work needs to address the role played by disorder and the absence of quasiparticle transport in these Cooper pair insulators. Further experiments that can establish the importance of Coulomb interactions and disorder to these SITs are also necessary.

Acknowledgements

This work was supported by the NSF through Grant No. DMR-0605797 and No. DMR-0907357, by the AFRL, and by the ONR. We are also grateful for support from AAUW. We thank our collaborators who contributed to the experiments highlighted in this manuscript: M. D. Stewart, Jr., H. Q. Nguyen, J. Shainline, E. Rudisaile, and Aijun Yin. Finally, we are grateful for discussions with N. Trivedi, I. Beloborodov, Y. Meir, M. Muller, and Z. Ovadyahu.

References

[1] G. Sambandamurthy, L. W. Engel, A. Johansson, and D. Shahar. Superconductivity-related insulating behavior. *Phys. Rev. Lett.*, 92(10):107005, Jan 2004.

[2] M. Steiner and A. Kapitulnik. Superconductivity in the insulating phase above the field-tuned superconductor-insulator transition in disordered indium oxide films. *Physica C*, 2005.

[3] V. F. Gantmakher, M. V. Golubkov, V. T. Dolgopolov, G. E. Tsydynzhapov, and A. A. Shashkin. Destruction of localized electron pairs above the magnetic-field-driven superconductor-insulator transition in amorphous in-o films. *JETP Lett.*, 68(4):363–369, Jan 1998.

[4] T. I. Baturina, A. Y. Mironov, V. M. Vinokur, M. R. Baklanov, and C. Strunk. Localized superconductivity in the quantum-critical region of the disorder-driven superconductor-insulator transition in tin thin films. *Phys. Rev. Lett.*, 99(25):257003, Dec 2007.

[5] Yen-Hsiang Lin and A Goldman. Magnetic-Field-Tuned Quantum Phase Transition in the Insulating Regime of Ultrathin Amorphous Bi Films. *Phys Rev Lett*, 106(12):1–4, March 2011.

[6] Y. Dubi, Y. Meir, and Y. Avishai. Theory of the magnetoresistance of disordered superconducting films. *Phys. Rev. B*, 73(5):054509, Jan 2006.

[7] V. M. Galitski, G. Refael, M. P. A. Fisher, and T. Senthil. Vortices and quasiparticles near the superconductor-insulator transition in thin films. *Phys. Rev. Lett.*, 95(7):077002, Jan 2005.

[8] S.M. Hollen, H.Q. Nguyen, E. Rudisaile, M.D. Stewart, J. Shainline, J.M. Xu, and J. M. Valles Jr. Cooper-pair insulator phase in superconducting amorphous Bi films induced by nanometer-scale thickness variations. *Phys. Rev. B*, 84(6), August 2011.

[9] M. D. Stewart Jr., A. Yin, J. M. Xu, and J. M. Valles Jr. Superconducting pair correlations in an amorphous insulating nanohoneycomb film. *Science*, 318(5854):1273–1275, Jan 2007.

[10] H. Q Nguyen, S. M. Hollen, M. D. Stewart, J. Shainline, Aijun Yin, J. M. Xu, and James M. Valles Jr. Observation of giant positive magnetoresistance in a cooper pair insulator. *Phys. Rev. Lett.*, 103(15):157001, Jan 2009.

[11] S. P. Benz, M. G. Forrester, M Tinkham, and C. J. Lobb. Positional disorder in superconducting wire networks and josephson junction arrays. *Phys. Rev. B*, 38(4):2869–2872, Jan 1988.

[12] J. A. Chervenak. *Studies of the Two Dimensional Superconductor Insulator Transition in Quench Condensed Films.* PhD thesis, Brown University, Providence, RI, 1998.

[13] E. Bielejec, J. Ruan, and W. Wu. Hard correlation gap observed in quench-condensed ultrathin beryllium. *Phys. Rev. Lett.*, 87(3):036801, Jan 2001.

[14] M. Müller. Purely electronic transport and localization in the Bose glass. *Ann. Phys.*, 18(12):849–855, December 2009.

[15] Karim Bouadim, Yen Lee Loh, Mohit Randeria, and Nandini Trivedi. Single- and two-particle energy gaps across the disorder-driven superconductor-insulator transition. *Nat Phys*, 7(7):1–6, July 2011.

[16] M. Müller. Private communication, 2011.

[17] I. Beloborodov, A. Lopatin, and V. Vinokur. Granular electronic systems. *Rev. Mod. Phys.*, 2007.

[18] A Ghosal, M Randeria, and N Trivedi. Role of spatial amplitude fluctuations in highly disordered s-wave superconductors. *Phys. Rev. Lett.*, 81(18):3940–3943, Jan 1998.

Journal of Physics: Conference Series **376** (2012) 012003 doi:10.1088/1742-6596/376/1/012003

Wonderful life at weak Coulomb interaction: increasing of superconducting/superfluid transition temperature by disorder

V.E.Kravtsov

Abdus Salam ICTP, P.O.Box 586, 34100 Trieste, Italy

E-mail: kravtsov@ictp.it

Abstract. We have shown that in systems where the Coulomb interaction is strongly suppressed, the superconducting transition temperature may be enhanced by disorder close to the Anderson localization transition. This phenomenon is based on the enhancement by disorder of the matrix element of attraction in the Cooper channel. For enhancement to take place one needs (i) strong disorder which makes the single-particle wave functions strongly inhomogeneous in space and (ii) strong correlation of the patterns of inhomogeneity for different wavefunctions. One case where such correlation is well known is the system close to the Anderson transition. We review the notion of multifractality of wavefunctions in this region and show how the enhancement of T_c arises out of the multifractal correlations.

Journal of Physics: Conference Series **376** (2012) 012003 doi:10.1088/1742-6596/376/1/012003

1. Introduction

The statement which is known under the name of the "Anderson theorem" and belongs to P.W.Anderson [1] and Abrikosov and Gorkov [2], reads that the non-magnetic disorder does not change the superconducting transition temperature T_c. As a matter of fact this statement has a status of a theorem based on the normalization and completeness of the set of single-particle wave functions only if the paring amplitude Δ does not have any variations in space or the latter can be averaged independently of the fluctuations of the single-particle wave functions. This is not true for sufficiently strong disorder where the joint effect of disorder and Coulomb interaction leads to suppression of T_c. This effect studied in detail by Finkelstein [3] was actually predicted earlier [4] as the leading correction to the mean field transition temperature in 2D disordered superconductors:

$$\frac{\delta T_c}{T_c} = -\frac{\lambda_{\rm eff}}{3g}\ln^3\left(\frac{\omega_0}{T_c}\right), \quad g = 2\pi^2\nu_0 D. \tag{1}$$

where $\lambda_{\rm eff} = \nu_0 V$ is the dimensionless constant of local interaction $V_q = V$, g is the dimensionless conductance, ν_0 is the DoS at the Fermi level, D is the diffusion constant and ω_0 is the Debye frequency.

For the screened Coulomb interaction in the universal limit $\lambda_{\rm eff} = 1$ and Eq.(1) corresponds to suppression of T_c by disorder.

$$\left(\frac{\delta T_c}{T_c}\right)_{\rm Coulomb} = -\frac{1}{3g}\ln^3\left(\frac{\omega_0}{T_c}\right), \tag{2}$$

However, if one assumes that the Coulomb interaction is absent (or strongly suppressed) and the only interaction that remains is the attraction in the Cooper channel characterizing by the small dimensionless constant $\lambda_{\rm eff} = -\lambda$, the same Eq.(1) together with the BCS relation $\lambda^{-1} = \ln(\omega_0/T_c)$ results in:

$$\left(\frac{\delta T_c}{T_c}\right)_{\rm Cooper} = +\frac{1}{3g}\ln^2\left(\frac{\omega_0}{T_c}\right), \tag{3}$$

Thus in the absence of Coulomb interaction one obtains the enhancement (w.r.t. the BCS result) of T_c by disorder by the same token as the suppression in the universal limit of the screened Coulomb interaction known as the Finkelstein effect [3].

It was not understood in the eighties that the origin of both suppression and enhancement is the weak multifractality of single-particle wave functions in 2D metals. The main goal of this paper is to make this connection physically transparent.

2. What is the multifractality?

While the earlier period of Anderson localization was mostly devoted to study of the scaling behavior of the localization or correlation length as a function of proximity to the mobility edge $E - E_c$ or to critical disorder $W - W_c$, the notion of multifractality of random critical single-particle wave function $\psi_n(\mathbf{r})$ was introduced by Wegner [5]

Journal of Physics: Conference Series **376** (2012) 012003 doi:10.1088/1742-6596/376/1/012003

almost immediately after the formulation of the scaling theory of Anderson transition. The definition was given in terms of the inverse participation ratio and its moments:

$$\int d^d\mathbf{r}\,\langle|\psi_n(\mathbf{r})|^{2q}\rangle \propto L^{-(q-1)\,d_q}, \tag{4}$$

where $\langle...\rangle$ denotes disorder average, L is the size of the system and d_q are the *fractal dimensions*. Eq.(4) covers all the three possible phases. The metal phase is characterized by the wave functions which occupy all the available volume with $|\psi_n(\mathbf{r})|^2 \sim L^{-d}$ by normalization. In this case one immediately concludes from Eq.(4) that all d_q are equal to the space dimensionality d. In the Anderson insulator $|\psi_n(\mathbf{r})|^2$ is of the order of $|\psi_n(\mathbf{r})|^2 \sim \xi^{-d}$ with the probability $(\xi/L)^d$ for a point $\mathbf{r}$ to fall inside the localization volume ξ^d, and is very small otherwise with the overwhelming probability $1-(\xi/L)^d$. Thus the leading term in Eq.(4) at large L is of the order of $L^d\xi^{-qd}(\xi/L)^d \sim \xi^{-(q-1)d}$ and hence it is independent of L, which formally corresponds to all $d_q = 0$. In the critical region close to the Anderson transition the set of critical exponents d_q is the an important and non-trivial characteristic of this transition, in addition to the well known exponent ν of the localization length $\xi \propto |W-W_c|^{-\nu}$.

Another way to characterize multifractality is to define the *spectrum of fractal dimensions* $f(\alpha)$ so that the volume in space V_α where the amplitude $|\psi_n(\mathbf{r})|^2$ scales like $L^{-\alpha}$ is $V_\alpha \propto L^{f(\alpha)}$. This approach explicitly takes into account the main feature of multifractality: the hierarchy of regions in space with different scaling of $|\psi|^2$ with the system size, each region being a fractal of the Hausdorf dimension $f(\alpha)$. With this definition we have for the integral of Eq.(4):

$$L^{-(q-1)d_q} \propto \int_0^\infty d\alpha\, L^{-q\alpha+f(\alpha)}. \tag{5}$$

By doing this integral in the saddle-point approximation valid at large $\ln L$ one concludes that the quantity $\tau_q = (q-1)\,d_q$ and $f(\alpha)$ are related by the Legendre transformation:

$$\tau_q = \alpha\, q - f(\alpha), \qquad \frac{df(\alpha)}{d\alpha} = q. \tag{6}$$

The validity of scaling Eq.(4) at the Anderson transition point was demonstrated in numerous numerical studies and analytically by $\epsilon = d-2$ expansion [5] of the nonlinear sigma-model (weak multifractality $d-d_q << d$) or by the virial expansion method [6] (strong multifractality $d_q << d$).

For further discussion it is very important to know the correlation of *two* single particle wavefunctions with the energies $E_n = E$ and $E_m = E+\omega$ which determines the matrix element of the local interaction. At the Anderson transition point this matrix element appears to be a power-law function of the energy difference ω as was conjectured by Chalker [7], checked by direct diagonalization of the lattice Anderson models and critical random matrix ensembles [8] and finally proven analytically for the critical random matrix ensemble [9]:

$$M_\omega = L^d \int d^d\mathbf{r}\,\langle|\psi_E(\mathbf{r})|^2\,|\psi_{E+\omega}(\mathbf{r})|^2\rangle \sim \left(\frac{E_0}{\omega}\right)^\mu, \tag{7}$$

where the exponent μ is closely related [7] with the fractal dimension d_2:

$$\mu = 1 - \frac{d_2}{d}. \tag{8}$$

The scaling Eq.(7),(8) in the frequency domain reflects the critical nature of single-particle states near the Anderson transition. However, it is valid well outside the mobility edge. In Fig.1 it is shown the result of calculation of the correlation function

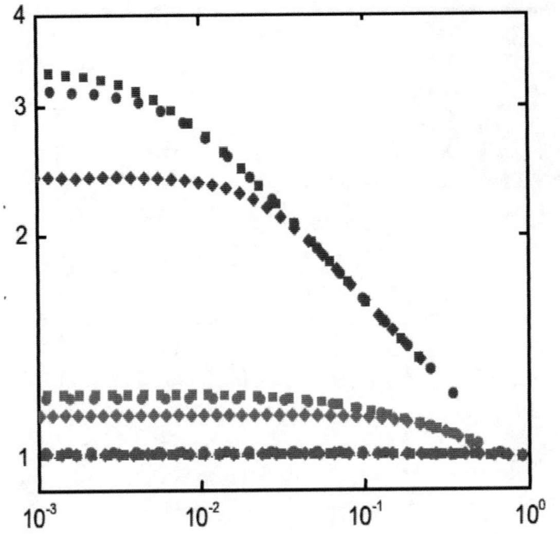

Figure 1. (color online) The correlation function Eq.(7) as a function of ω from direct diagonalization of the 3D Anderson model with critical disorder W=16.5 for disorder strength W=2 (green), W=5 (red) and W=10 (blue), and different system sizes L=8(diamond),L=16 (circles),L=20 (squares). At $W = 10$ and $L = 16, 20$ the critical power-law is well seen for $(\nu_0 \xi^d)^{-1} = \delta_\xi < \omega < E_0$.

Eq.(7) by direct diagonalization of the 3D Anderson model in the metallic phase with sub-critical disorder. One can see that even relatively far from the critical disorder ($W_c = 16.5$), at $W = 10$, the power-law critical behavior is well seen. It is saturated when the ω-dependent "resolution length" $L_\omega = (\nu_0 \omega)^{-1/d}$ becomes larger than either the correlation length $\xi \propto (W_c - W)^{-\nu}$ or the system size L. Even at disorder as small as $W = 5$ the enhancement of M_ω given by the factor $(E_0/\omega)^\mu$ in Eq.(7) is quite significant. Only at $W = 2$ when $\xi < \ell = (\nu_0 E_0)^{\frac{1}{d}}$ it disappears, and we reach the limit $L^d M_\omega = 1$ expected for the extended single particle wavefunctions which occupy all the available space.

Such a behavior of the correlation function Eq.(7) suggests a stricture of the typical single-particle wavefunction at the sub-critical disorder which is sketched in Fig.2. It consists of the "fractal regions" of the size of the "correlation radius" ξ appearing in the scaling theory of localization. Inside such regions all the correlations are identical to those in the critical wavefunction. As the energy difference ω decreases the "resolution length" $L_\omega = (\nu_0 \omega)^{-\frac{1}{d}}$ increases and at $\omega < \delta_\xi = (\nu_0 \xi^d)^{-1}$ it exceeds the correlation

Journal of Physics: Conference Series **376** (2012) 012003 doi:10.1088/1742-6596/376/1/012003

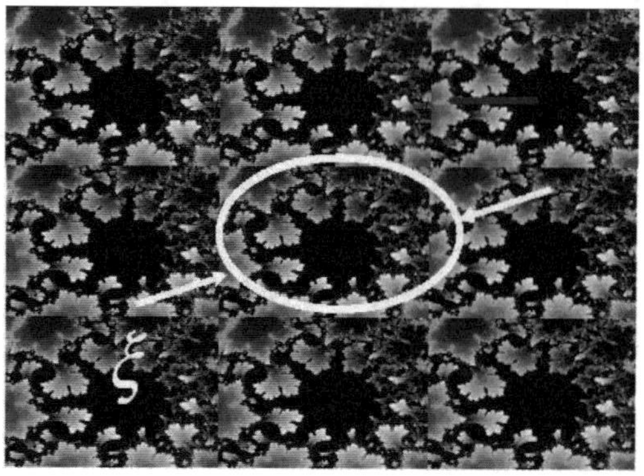

Figure 2. (color online) The cartoon of the amplitude landscape of the sub-critical single-particle wavefunction. The correlation length $\xi \propto (W_c - W)^{-\nu}$ sets the typical size of the "fractal region" where the wavefunction correlations are the same as in the critical state. If the ω-dependent "resolution length" $L_\omega = (\nu_0\omega)^{-\frac{1}{d}}$ is smaller than ξ the critical power law in Eq.(7) is well seen. For larger resolution length one observes a saturation in the ω-dependence which is typical of metal. However, the fact that there are many peaks (dark regions on the plot)in the wavefunction landscape where the amplitude is pretty high, leads to the enhanced saturation level. The exception is the case of a very weak disorder when the correlation length $\xi \ll \ell$ is smaller than the size $\ell = (\nu_0 E_0)^{-\frac{1}{d}}$ of the "pixel" which is only slightly larger than the atomic scale. Only in this limit the Anderson theorem may apply. The similar cartoon holds for the localized states at super-critical disorder. In this case the amplitude is exponentially small outside the localization radius ξ but has the same fractal structure inside this radius. That allows to speak about the "multifractal metal" and "multifractal insulator" [8].

length ξ. Only for such ω the correlation function Eq.(7) may show that the state is not critical but extended. Thus $\omega = \delta_\xi$ marks the onset of saturation in Fig.1 for $L \gg \xi$.

It is important that at large $\omega = E_0$ all curves in Fig.1 merge, in agreement with Eq.(7). This means that for $\omega < E_0$ the matrix element M_ω is *enhanced* compared to the ideal case of totally extended wavefunctions which occupy all the available space without peaks and holes at a scale much larger than the Fermi wavelength. In order to better understand the origin of this enhancement we emphasize that it is caused by the *inhomogeneity* of the amplitude $|\psi(\mathbf{r})|^2$ in space. The presence of "holes" with very small amplitude in certain regions implies appearance of peaks in other regions, as the integral of the amplitude over entire system of the size L is equal to 1 by normalization. The key point is that the positions of the peaks in two different wave functions are highly correlated for the critical and slightly off-critical states. Contrary to naive expectation, even at large energy difference ω which is only slightly smaller than the *upper energy scale for fractality* E_0 (for the 3D Anderson model it is approximately 1/3 of the bandwidth), the correlation is pretty high. This leads to the very slow decay of the matrix element $M_\omega \sim \omega^{-\mu}$ ($\mu < 1$) with increasing ω [10, 11]. Note that the ultimate reason for such a correlation is the presence of valleys in the random potential. All the wave functions are the solutions to the Schroedinger equation in the *same* random potential. So, the peaks know where they predominantly want to be situated. Yet, the criticality matters, as for localized states the similar correlation is absent, perhaps because there are too many possibilities to arrange a localized state of small localization radius. This is the reason why Eq.(7) can be formally extended to ideal metal by setting $d_2 = d$ but it cannot be extended to hard insulator which formally corresponds to $d_2 = 0$.

3. The minimal model in real and in Fock space

As has been already mentioned in Introduction, in the situation when Coulomb interaction can be neglected the minimal model consists of the single-particle Hamiltonian with disorder and the local attractive interaction [12]:

$$\hat{H} = \sum_{\mathbf{r},\mathbf{r}',\alpha} \varepsilon_{\mathbf{r},\mathbf{r}'}\, \Psi^\dagger_\alpha(\mathbf{r})\Psi_\alpha(\mathbf{r}') - U\sum_{\mathbf{r}} \Psi^\dagger_\uparrow(\mathbf{r})\Psi^\dagger_\downarrow(\mathbf{r})\Psi_\downarrow(\mathbf{r})\Psi_\uparrow(\mathbf{r}). \tag{9}$$

In the case where disorder is strong it is very useful to switch from the coordinate representation to the representation of the exact single-particle states where all effect of disorder is taken into account [13, 14]. This is done as usual by substitution $\Psi_\alpha(\mathbf{r}) = \sum_n \psi_n(\mathbf{r})\, c_{n,\alpha}$ of the Ψ-operator in terms of the exact single-particle wavefunctions $\psi_n(\mathbf{r})$ and the creation/annihilation operators $c^\dagger_{n,\alpha}$ and $c_{n,\alpha}$ of electrons in the corresponding state n with the spin α. As the result the single-particle part of the Hamiltonian becomes diagonal but the interaction part produces a variety of terms

$$\hat{H} = \sum_{n,\alpha} E_n\, c^\dagger_{n,\alpha}c_{n,\alpha} - U\sum_{n,k,m,l} M_{nkml}c^\dagger_{n,\uparrow}c^\dagger_{k,\downarrow}c_{m,\downarrow}c_{l,\uparrow}, \tag{10}$$

where

$$M_{nkml} = \int d^d\mathbf{r}\, \psi^*_n(\mathbf{r})\psi^*_k(\mathbf{r})\psi_m(\mathbf{r})\psi_l(\mathbf{r}). \tag{11}$$

These terms may be divided into two major groups. The first one consists of the terms where all four indices are the same $n = k = m = l$ or there are two pairs of equal indices. The examples are $M_n \equiv M_{nnnn}$ and $M_{nm} \equiv M_{nnmm}$, as well as $\bar{M}_{nm} \equiv M_{nmmn}$. They do not change the parity of occupation of a single particle state but can only move a singlet pair from state n to state m (M_{nnmm}) or to flip the spin of the singly occupied state (M_{nmmn}). All the other terms may convert the singly-occupied state into the empty or doubly occupied states and vise versa.

In the region of localized single-particle states the largest matrix element is M_n, all the other matrix elements are smaller because different single-particle states rarely overlap. When the matrix element $-UM_n$ is large enough, it favors a *local pair formation* by producing a gap between the *even* many-body states in which all the single-particle states are occupied by a singlet pair or empty, and all other (*odd*) states. In this case one may neglect the presence of odd states whatsoever and introduce bosonic operators

$b_n^\dagger = c_{n\uparrow}^\dagger c_{n\downarrow}^\dagger$. One can easily check that $(b_n^\dagger)^2 = 0$, so that we introduced a *hard-core* bosons. In addition to that, for even states $b_n^\dagger b_n = c_{n\uparrow}^\dagger c_{n\uparrow} = c_{n\downarrow}^\dagger c_{n\downarrow}$.

The *minimal model in the Fock space* is obtained by neglecting of all terms except for those proportional to M_n and M_{nm}. The corresponding Hamiltonian takes the form:

$$\hat{H} = \sum_n 2E_n\, b_n^\dagger b_n - U \sum_{n,m} M_{nm}\, b_n^\dagger b_m + h.c. \tag{12}$$

It describes physics of local pairs with hard-core interaction attached to single-particle states n with random energies $2E_n$ and hopping due to the attractive interaction of original electrons. The term proportional to M_n is included into the chemical potential. Obviously, the hopping favors establishing a delocalized phase which at low enough temperature must be superfluid. Disorder, in contrast, is trying to localize the pairs.

Eq.(12) admits also a spin representation in which the hard-core nature of bosons is automatically included. To this end one introduces the spin-1/2 operators $S_n^+ = b_n^\dagger = c_{n\uparrow}^\dagger c_{n\downarrow}^\dagger$ and $S_n^z = b_n^\dagger b_n - \frac{1}{2} = \frac{1}{2}\,[c_{n\uparrow}^\dagger c_{n\uparrow} + c_{n\downarrow}^\dagger c_{n\downarrow} - 1]$. Then Eq.(12) takes the form:

$$\hat{H} = \sum_n 2E_n\, S_n^z - U \sum_{n,m} M_{nm}\, S_n^+ S_m^- + h.c. \tag{13}$$

In this language the Ising order corresponds to localized pairs and the XY order corresponds to a superfluid phase.

So far we neglected the term proportional to M_{nmmn} which is also operating in the even sector of the Hilbert space. It can be easily taken into account by adding the following term to the Hamiltonian:

$$-U \sum_{n,m} \bar{M}_{nm}\, S_n^z S_m^z = -U \sum_{n,m} \bar{M}_{nm}\, b_n^\dagger b_n b_m^\dagger b_m + UV^{-1} \sum_n b_n^\dagger b_n. \tag{14}$$

This term describes interaction (attraction) of bosons at different points in the Fock space. It favors phase separation with fixed pair occupation number 1 or 0 and thus disfavors establishing a coherent superfluid phase where the occupation number strongly fluctuates. This has an effect of reducing the superconducting/superfluid transition temperature. Such a suppression of T_c by a factor of order 1 [16] does not change, however, the main conclusions of this paper.

Among other omitted terms the most significant are the following:

$$-U \sum_{n,m,l} M_{nnml}\, b_n^\dagger c_{m\downarrow} c_{l\uparrow} + h.c. \tag{15}$$

which describe dissociation and creation of pairs out of singly occupied single-particle states. They can be safely neglected in the region of sufficiently strongly localized single-particle states but have to be taken into account in the region of critical and sub-critical states where they determine the Ginzburg number Gi.

4. Mean field approximation in the Fock space

The minimal model Eq.(13) is a good starting point not only in the region of localized single-particle states (*the "pseudo-gap" region*) but also in the critical and sub-critical region, although in this case its derivation from the minimal model in real space Eq.(9) is not justified by a small parameter. Indeed, the standard mean-field treatment of the spin-model Eq.(13) yields the following equation for the critical temperature [15]:

$$\Delta_E = \frac{\lambda}{2}\int dE'\,\Delta_{E'}\,\frac{\tanh\left(\frac{E'}{T}\right)}{E'}\,M_{E-E'}, \tag{16}$$

where M_ω is given by Eq.(7) and $\lambda = U\nu_0$ is the dimensionless attraction constant.

At a weak disorder $M_\omega = 1$ for $|\omega| < \omega_D$ and zero otherwise, and one can immediately recognize the standard BCS equation for T_c, albeit with T instead of $2T$. This is the price of restriction to the odd sector of the Hilbert space. In case when there is no gap between the many-body states of the even and odd sectors, the latter is thermodynamically relevant even if dynamically both sectors are totally decoupled. We will show, however, that the accuracy of the mean-field approximation on the metallic side close to the Anderson localization transition is up to a constant pre-factor of order 1. With this uncertainty the difference by a factor of 2 discussed above is beyond the accuracy of the mean-field approach.

Let us apply the mean-field equation (16) to compute the T_c in the critical region. Plugging Eq.(7) in Eq.(16) and making a rescaling $E \to T_c\,\varepsilon$, $E' \to T_c\,\varepsilon'$ one reduces the problem of T_c to that of the eigenvalue $\bar{\lambda}^{-1} \equiv \lambda^{-1}\,(T_c/E_0)^{\mu} \sim 1$ of the dimensionless linear integral operator. Then we immediately conclude that:

$$T_c \sim E_0\,\lambda^{\frac{1}{\mu}}, \quad \mu = 1 - \frac{d_2}{d}. \tag{17}$$

5. Ginzburg number and the accuracy of the mean-field approximation

As is well known, the thermodynamic fluctuations of phase and fluctuation of the local T_c due to disorder restrict the region of validity of the mean-field approximation. To take into account correctly the first effect in the region of *extended* single-particle states one needs to include dissociation processes Eq.(15). Without such processes, for the minimum model in the Fock space Eq.(13), the mean-field approximation is exact in this region, as the matrix element M_{nm} couples to an infinite number of states in the thermodynamic limit. When dissociation processes are properly accounted for [16] one arrives at a remarkable result:

$$\mathrm{Gi} \equiv \frac{\Delta T_c}{T_c} \sim 1, \tag{18}$$

where ΔT_c is the "fluctuation region". The effects of multifractality cancel out in the Gi number, and it appears to be a universal number of order 1 close to the Anderson localization transition [14].

In the region of sufficiently strongly localized single-particle states, the Gi number is mainly controlled by the effective coordination number K of states coupled by the matrix element M_{nm} in the minimal spin model Eq.(13). As one goes away from the Anderson transition, this number decreases. One can show [16] that in this region:

$$\mathrm{Gi} = \mathrm{Gi}_1 + \frac{\mathrm{Gi}_2}{K^2}, \tag{19}$$

where Gi_1 and Gi_2 are universal numbers of order 1. This effect of decreasing of coordination number with increasing disorder invalidates the mean-field approximation at sufficiently strong disorder and finally leads to the superconductor to insulator transition.

Note however, that in the vicinity of the Anderson transition Eq.(18) holds true. Given the relationship between the Ginzburg number and the energy scale J associated with the phase rigidity:

$$J = \frac{T_c^{(MF)}}{\mathrm{Gi}^{\frac{1}{3}}} \tag{20}$$

one concludes that the scale J and the mean-field transition temperature $T_c^{(MF)}$ given by Eq.(17) differ by a universal factor of order 1. This implies that the phase fluctuations may reduce the true superconducting transition temperature by at most a universal factor of order 1 compared to the mean-field result Eq.(17).

6. Virial expansion method

In order to describe the behavior of transition temperature in the region of localized single-particle states where the mean-field approach breaks down, we exploit [16] the idea of virial expansion applied to the spin-model Eq.(13). To this end we analytically express the Cooper susceptibility $\chi_C(T)$ (the response of the spin S^+ to the infinitesimal perturbation $\Delta\, S^+$) as a sum of contributions of clusters of M coupled spins:

$$\chi_C(T) = \sum_{M+1}^{\infty} \chi_M(T), \tag{21}$$

The true transition temperature corresponds to temperature T_c when this series becomes divergent. According to D'Alembert criterion this happens when:

$$\lim_{M\to\infty} \frac{\chi_{M+1}(T_c)}{\chi_M(T_c)} = 1. \tag{22}$$

In view of rapid increase of complexity of the analytical expressions for χ_M with increasing M, we truncated the series Eq.(21) and used the operative definition of T_c in terms of small-cluster contributions $T_c^{(1)}$ and $T_c^{(2)}$:

$$\chi_1(T_c^{(1)}) = \chi_2(T_c^{(1)}), \quad \chi_2(T_c^{(2)}) = \chi_3(T_c^{(2)}). \tag{23}$$

By evaluating χ_1, χ_2 and χ_3 taking the energies E_n and matrix elements M_{nm} from exact diagonalization of the Anderson model on a 3D lattice and solving Eq.(23) numerically we were able to find statistics of $T_c^{(1)}$ and $T_c^{(2)}$ at different positions of Fermi level relative to the mobility edge. The average $T_c^{(1)}$ and $T_c^{(2)}$ appear to be in good agreement with each other which gave us a reasonable confidence in the convergence of the procedure. In Fig.3, the results of virial expansion calculation for the average $\langle T_c^{(2)}\rangle$ are presented as a function of the Fermi level position. It is remarkable that the truncated virial expansion method is in good agreement with the results of mean-field calculations in the region of extended and weakly localized states and gives substantially smaller transition

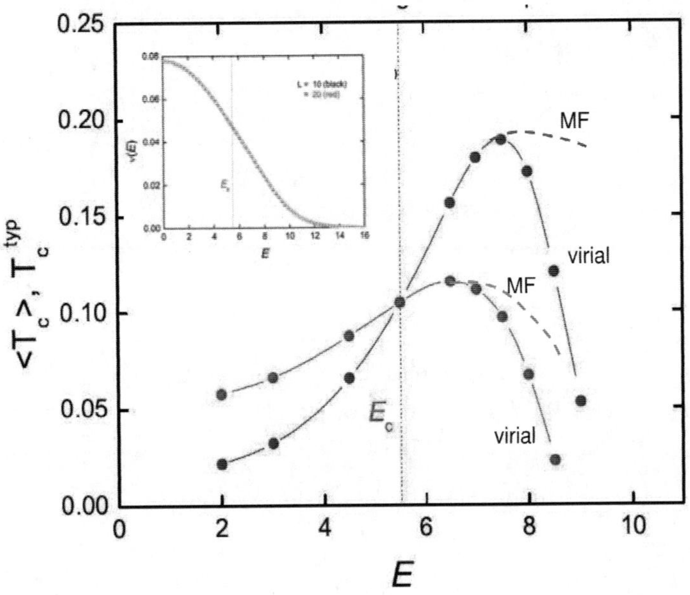

Figure 3. (color online)Transition temperature as a function of the Fermi energy position relative the mobility edge E_c for the fixed dimensionless constant $\lambda = 0.08$ (blue circles) and for the fixed attractive interaction $U = 1.7t$(green circles), where t is the hopping integral of the Anderson model. In the insert: energy dependence of the mean density of states. Large energy corresponds to region of localized single-particle states. Inside this region the mean-field (MF) transition temperature is substantially larger than that ($T_c \equiv \langle T_c^{(2)} \rangle$) found from the truncated virial expansion series. The Cooper susceptibilities $\chi_2(T)$ and $\chi_3(T)$ were calculated numerically using exact eigenfunctions and eigenvalues of the Anderson model with strength of disorder $W = 4$, size of system $L = 20^3$ and the Debye frequency $\omega_0 = 0.5$.

temperature for stronger localized single-particle states. This proves that the virial expansion method, even with only 3 terms retained in the series Eq.(22), captures the decrease of the effective coordination number K and is not equivalent to the mean-field approximation.

7. Enhancement of T_c near the Anderson transition.

Equation (17) shows that when the Fermi energy is at the mobility edge $E = E_c$ the critical temperature behaves as a power law of the dimensionless attraction constant λ. Numerical simulations on the 3D Anderson model of localization give (see Ref. [16] and references therein) the value of the fractal dimension $d_2 = 1.29 \pm 0.1$. Thus the exponent in the power law dependence of T_c on λ is close to $1/\mu \approx 1.8$. For small λ one always has $\lambda^{1.8} \gg \exp(1/\lambda)$. In addition to that the characteristic energy E_0 is of electronic nature and it is typically higher than the Debye frequency ω_0. Therefore we conclude that at small λ the transition temperature given by Eq.(17) is much higher than the BCS transition temperature $T_c \sim \omega_0 \exp(-1/\lambda)$. This enhancement of T_c is also obtained by a virial expansion method which is completely independent of the mean-field approach. It can be easily traced back to the enhancement of the matrix element M_{nm} shown in Fig.1.

In order to figure out how general is the effect of enhancement of T_c at negligible Coulomb interaction, let us consider the case of weak multifractality $1 - d_2/d \ll 1$. This case is relevant for the 2D metal [17, 18], where

$$1 - \frac{d_2}{2} = \frac{1}{g}, \quad g = 2\pi^2 \nu_0 D. \tag{24}$$

Expanding the matrix element *Eq.*(7) in the mean-field equation Eq.(16) up to the leading correction in $1/g$ $M_\omega \approx 1 + \frac{1}{g} \ln(E_0/\omega)$ we obtain the $1/g$ correction to the transition temperature:

$$\frac{\delta T_c}{T_c} = \frac{1}{2g} \ln\left(\frac{\omega_0}{T_c}\right) \ln\left(\frac{E_0^2}{\omega_0 T_c}\right) \sim \frac{1}{g} \ln^2\left(\frac{\omega_0}{T_c}\right). \tag{25}$$

We obtained the correction of the same order as in Ref.[4] given by Eq.(3) thus proving that our analysis at $\lambda \ll 1$ qualitatively applies also to the case of *weak multifractality* in 2D metals. Then Eq.(2) tells us that the really crucial assumption for enhancement of T_c by disorder is the suppression of Coulomb interaction. Recently [19] this statement was confirmed by the RG analysis of the Finkelstein-like nonlinear sigma-model in $d = 2$ and $d = 2 + \epsilon$ where only the short-range interactions were taken into account.

There are certain systems [20, 21] where one may suspect such a suppression. Moreover, the enhancement of T_c just before the onset of the insulating behavior was observed in one of them with the dependence of T_c on doping highly reminiscent of the Fig.3. More investigations are needed to confirm or reject this conjecture. In any case, the search for systems with suppressed Coulomb interaction (except for obvious case of cold neutral fermionic atoms with attractive interaction) is a challenging task which may open a door into a wonderful world without Coulomb interaction.

References

[1] Anderson P W, 1959 *J. Phys. Chem. Solids* **11** 26.
[2] Abrikosov A A, Gor'kov L P, 1958 *Zh. Eksp. Teor. Fiz.* **35** 1558; 1959 *Zh. Eksp. Teor. Fiz.* **38** 319 [*Sov. Phys. JETP* **8**, 1090 (1959); **9**, 220 (1959)]; Abrikosov A A, Gor'kov L P, 1961 *Zh. Eksp. Teor. Fiz.* **39** 1781 [*Sov. Phys. JETP* **12**,1243 (1961)].
[3] Finkelstein A M, 1987 *Pis'ma ZhETF* **45** 37 [*Sov.Phys.JETP Letters* **45** 46 (1987)]
[4] Maekawa S, Fukuyama H, 1982 *J.Phys.Soc.Jpn.* **51** 1380.
[5] Wegner F, 1980 *Z. Phys. B* **36** 209
[6] Levitov L S, 1990 *Phys.Rev.Lett.* **64** 547; Mirlin A D, Evers F, 2000 *Phys.Rev.B* **62** 7920; Yevtushenko O, Kravtsov V E, 2003 *J.Phys.A:Math.Gen.* **36** 8265; Yevtushenko O, Ossipov A, 2007 *J.Phys.A: Math.Gen.* **40** 4691
[7] Chalker J T, Daniel G J, 1988 *Phys.Rev.Lett.* **61** 593; Chalker J T, 1990 *Physica A* **167** 253
[8] Kravstov V E, Cuevas E, 2007 *Phys.Rev.B* **76** 235119
[9] Kravstov V E, Ossipov A, Yevtushenko O M, 2010 *Phys.Rev.B* **82** 161102

[10] Kravtsov V E, Muttalib K A, 1997 *Phys.Rev.Lett.* **79** 1913
[11] Mirlin A D, Fyodorov V Y, 1997 *Phys.Rev.B* **55** 16001
[12] Ghosal A, Randeria M, Trivedi N, 2001 *Phys.Rev.B* **65** 014501
[13] Ma M, Lee P A, 1985 *Phys.Rev.B* **32** 5658
[14] Kapitulnik A, Kotliar G, 1985 *Phys.Rev.Lett.* **54** 473; Kotliar G, Kapitulnik A, 1986 *Phys.Rev.B* **33** 3146
[15] Feigelman M V, Ioffe L B, Kravtsov V E, Yuzbashyan E A, 2007 *Phys.Rev.Lett.* **98** 027001
[16] Feigelman M V, Ioffe L B, Kravtsov V E, Cuevas E, 2010 *Ann. Phys.* **325** 1368
[17] Altshuler B L, Kravtsov V E, Lerner I V, 1986 *JETP Letters* **43** 441
[18] Efetov K B., Fal'ko V I, 1995 *Europhys. Lett.* **32** 627
[19] Burmistrov I S, Gornyi I V, Mirlin A D, 2011 arXiv:1102.3323.
[20] Yamanaka S, 2010 *J. Mater. Chem.* **20** 2922
[21] Taguchi Y, Kitora A, Iwasa Y, 2006 *Phys.Rev.Lett.* **97** 107001

Comparative Study of the Magnetic Susceptibility of n-Ge near the Insulator–Metal Phase Transition by Means of ESR and SQUID

A.I. Veinger, A.G. Zabrodskii, T.L. Makarova, T.V. Tisnek, S.I. Goloshchapov and P.V. Semenikhin

Ioffe Physico-Technical Institute of the Russian Academy of Sciences, Politekhnicheskaya 26, St.Petersburg, 194021, Russia

Impurity paramagnetic susceptibility of n-Ge has been measured in the vicinity of the insulator–metal phase transition by two techniques: ESR and SQUID. A comparison of these methods demonstrated that their results qualitatively coincide at low temperatures, but differ at high temperatures. The advantage of the ESR technique is in its substantially higher relative precision, but SQUID makes it possible to obtain on principle absolute values of the magnetic susceptibility. It was found that, in the present-day state of development, the SQUID method has a substantially poorer precision, compared with ESR

Keywords: Magnetic susceptibility, Insulator – metal phase transition, Semiconductor, Germanium
PACS: 75.20.Ck, 75.30.Cr

INTRODUCTION

It is well known that an increase in the impurity concentration in a semiconductor gives rise to interaction between separate impurity centers, as a result of which the semiconductor passes from the insulator to the metallic state. This problem, a part of the general problem of the insulator–metal (IM) phase transition in a solid, is widely discussed in the scientific literature and, in particular, in a number of monographs (see, e.g., [1, 2]). It is known, the magnetic properties of a substance change in the vicinity of the transition: the insulator state is characterized by Curie paramagnetism or antiferromagnetism, whereas in the metallic state, the magnetic properties of the substance are described by the Pauli paramagnetism or ferromagnetism.

In nonmagnetic semiconductors (e.g., in Ge and Si), the paramagnetism is determined by impurities. Each atom of a shallow impurity has an unpaired spin, and the ensemble of these spins is manifested as a paramagnetic contribution to the magnetic susceptibility χ. The temperature dependence of χ in the absence of a spin–spin interaction and at temperatures above 1 K is described by the Curie law:

$$\chi = n_S \mu_B^2 p_{ef}^2 / 3kT, \qquad (1)$$

Journal of Physics: Conference Series **376** (2012) 012004 doi:10.1088/1742-6596/376/1/012004

where n_S - is the spin concentration; μ_B - Bohr magneton; p_{eff}, - effective magnetic moment per atom (p_{eff}=1 for the s-state); k - Boltzmann constant; and T - temperature.

The interaction of shallow impurities near the IM phase transition gives rise to cooperative phenomena: spins may be oriented in parallel to each other (ferromagnetic interaction) or be antiparallel (antiferromagnetic interaction). Analysis of interactions of this kind furnishes information about specific features of these phase transitions. If there are spin interactions, the temperature dependences $\chi(T)$ above the Curie point (for ferromagnetics) or Neel point (for aniferromagnetics) obey the Curie–Weiss law:

$$\chi = n_S \mu_B^2 p_{ef}^2 / 3k(T+\Theta), \qquad (2)$$

where the parameters have the same meanings as those in (1), but a temperature constant Θ is added. This constant is positive for a ferromagnetic and negative for an antiferromagnetic.

It is follows from (2) that, if the dependence $1/\chi = f(T)$ is plotted, it crosses the abscissa axis, making an intercept Θ, with the interaction being ferromagnetic at low temperatures at $\Theta > 0$ and antiferromagnetic at $\Theta < 0$.

The most effective way to study the magnetic properties of shallow impurities in a semiconductor is by the ESR technique [3]. In this case, a model material is silicon. All stages of variation of ESR spectra with increasing spin density, caused by the spin–spin interaction, have been observed for this semiconductor doped with phosphorus [4].

However, these measurements were insufficiently detailed because of the small number of samples in the set under study and insufficiently gradual change in the spin density. A more detailed study could be performed for this range of spin concentrations in Ge:As samples [5, 6].

A study of the ESR in n-Ge:As in the vicinity of the IM phase transition makes it possible to obtain an abundant information about changes in the magnetic properties of the impurity subsystem in this range of concentrations. However, this technique has an important disadvantage which consists in that it is by no means applicable for all paramagnetic impurities. The most interesting impurities in this regard are acceptors in Ge. They are also paramagnetic, but commonly do not appear in ESR because of the another type of the hole wave function [3].

The magnetic properties of p-Ge in the vicinity of the IM phase transition can be studied by measuring the static magnetic susceptibility. But it is necessary to find out how the results are precise that can be obtained from measurements of this kind. We made an attempt to solve this problem by comparing the results of measurements of the static magnetic susceptibility (χ_{stat}) of n-Ge:As on a SQUID installation and the dynamic magnetic susceptibility (χ_{dyn}) by means of ESR. An analysis of results obtained using both techniques for n-Ge:As is the goal of the present study.

COMPARISON OF THE RESULTS OF THE PARAMAGNETIC SUSCEPTIBILITY MEASUREMENTS IN N-GE:AS

The magnetic susceptibility was measured on the samples previously used to study ESR spectra [5, 6]. It was found from the resonance ESR absorption signal (dynamic magnetic susceptibility χ) as $\chi \approx \int PdH$, with the existence of a skin layer, changes in the Q-factor of the resonator, and the Dysonian shape of the absorption line taken into account. The static magnetic susceptibility was measured with a SQUID MPMS-XL-1 instrument operating in the temperature range 1.7–400 K in magnetic fields of up to 10 kOe. The "reciprocal sample option" function improved the measurement accuracy to 10^{-9} cm^3/g.

Figure 1 shows temperature dependences of the dynamic magnetic susceptibility for a set of samples with spin concentrations close to the critical concentration for the IM phase transition ($n_C = 3.7\times10^{17}$ cm^{-3}). The same figure presents, for comparison, the dependence corresponding to the Curie law. It follows from (1) that the slope ratio of the curve obeys this law when the number of paramagnetic species is preserved as the temperature is varied; a faster decrease with increasing temperature occurs if the number of paramagnetic species becomes smaller, and a slower decrease, if this number grows.

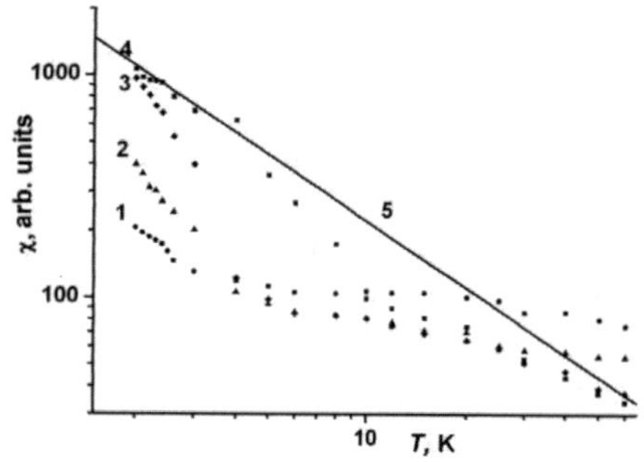

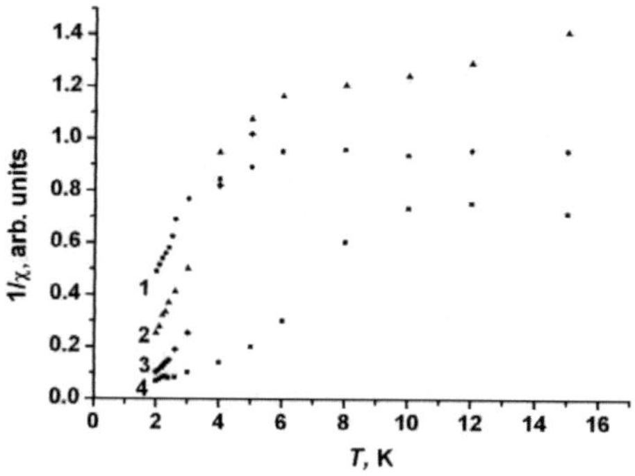

FIGURE 1. Temperature dependences of the dynamic paramagnetic susceptibility for samples with various n/n_C ratios: (1) 0.97, (2) 0.77, (3) 0.53, (4) 0.39 and compensation K: (1) 0, (2) 0.28, (3) 0.56, (4) 0.68, (5) Curie law.

FIGURE 2. Temperature dependences of the inverse dynamic magnetic susceptibility for the same samples.

All the temperature dependences show three regions of departure from the Curie law: low-temperature ($T \leq 5K$), in which χ rapidly decreases with increasing temperature; intermediate ($5\ K \leq T \leq 20\ K$), in which χ is nearly invariable; and high-temperature ($T \geq 20\ K$), in which χ gradually decreases.

Plotting the dependence $1/\chi(T)$ makes it possible to determine the nature of interaction between spins at their different densities near the IM phase transition. These dependences are shown in Fig. 2 for several samples for the low-temperature range ($T < 15\ K$). The intersection of the dependences in the figure with the abscissa axis can be used to judge, in accordance with (2), about the nature of the carrier interaction. At high temperatures ($T > 6\ K$), $1/\chi(T)$ gradually decreases as the temperature becomes lower. This temperature range was analyzed in our previous studies [7, 8]. The behavior of the magnetic susceptibility at these temperatures was attributed to the appearance of antiferromagnetically coupled pairs of spins. However, using now the digital signal recording and a cryostat working at temperatures down to 2 K, we could reveal additional low-temperature region.

In this range χ rapidly grows with the temperature falls and, consequently, the function $1/\chi(T)$ rapidly decreases. As can be seen from these curves, the decrease of the function $1/\chi(T)$ below 3.5 K can be represented by a straight line crossing the abscissa axis at $T > 0$. In this case, it should be admitted that, in accordance with the Curie–Weiss law, a ferromagnetic coupling of spins is observed in this temperature range. With approach to the critical point, this effect becomes less pronounced and sample 1, the closest to the transition, is a paramagnetic and the dependence $1/\chi(T)$ for this samples has $\Theta = 0$.

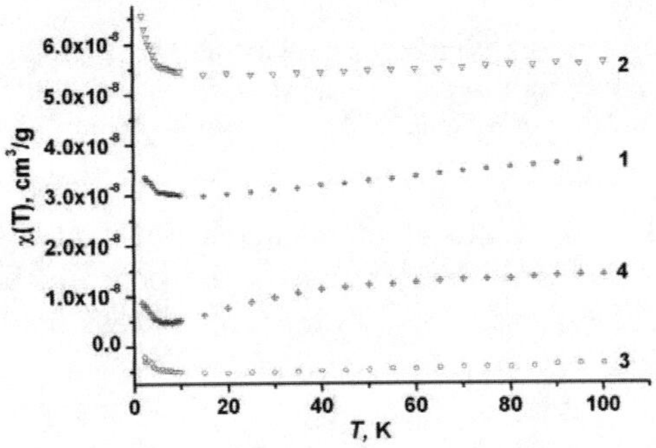

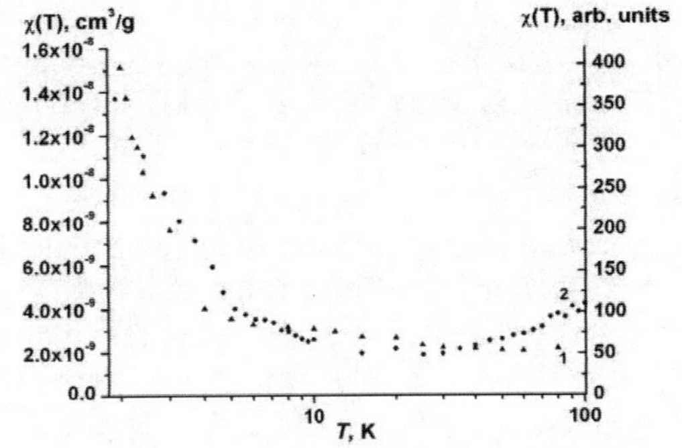

FIGURE 3. Temperature dependences of the static magnetic susceptibility for the same samples after subtracting the lattice susceptibility.

FIGURE 4. Superimposed temperature dependences of χ, determined from EPR and SQUID: (1) EPR, right-hand scale; and (2) SQUID, left-hand scale (sample 2).

Our experiments demonstrated that all the samples under study are diamagnetic. This is natural because the lattice diamagnetic susceptibility substantially exceeds the impurity paramagnetic susceptibility. After its diamagnetic part is subtracted from the full susceptibility, the temperature dependences $\chi(T)$ take the form shown in Fig.3. It can be seen that the temperature-dependent part of χ has two characteristic portions: an increase in temperature at low temperatures ($T \leq 10$ K) leads to a decrease in χ, and that at high temperatures ($T \geq 10$ K), to an increase in this parameter. Comparison with Fig.1 shows that the behavior of the temperature-dependent part of the curves at low temperatures qualitatively coincides with the results of EPR measurements, i.e., it is determined by the presence of impurity spins in a sample. However, an increase in temperature in ESR measurements leads to a substantial decrease in χ, whereas in SQUID measurements, this decrease is less than 20%. This difference is indicative of a large contribution to the magnetic susceptibility from other materials present in the measuring volume and characterized by a magnetic moment proportional to the field and a temperature-independent χ.

To combine the dependences $\chi(T)$ obtained by using both techniques, we transformed the static susceptibility so that these dependences coincided in the low-temperature range. The result of such transformation for one of our samples is presented in Fig. 4.

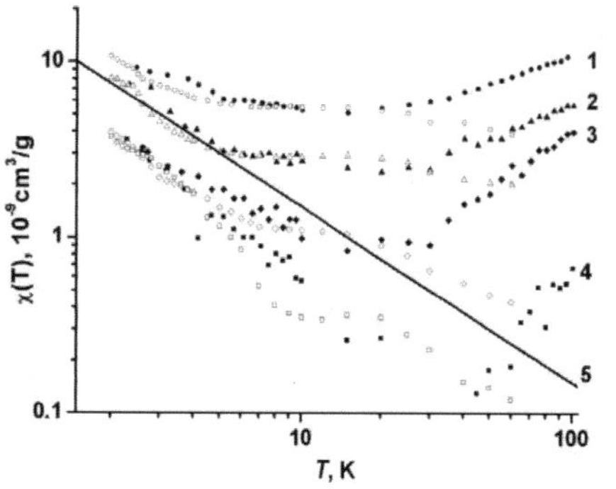

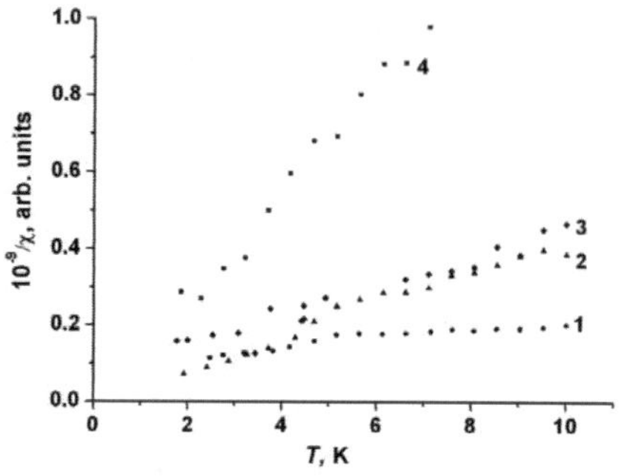

FIGURE 5. Temperature dependences of the static (black symbols) and dynamic (white symbols) magnetic susceptibility χ for the same n-Ge:As samples upon fitting; 5 – Curie law.

FIGURE 6. Temperature dependences of the inverse static magnetic susceptibility for the same samples upon fitting.

Comparison of the temperature dependences of the dynamic and static magnetic susceptibilities shows that the most noticeable difference between the dependences $\chi(T)$ appears at high temperatures: $\chi_{dyn}(T)$ decreases and $\chi_{stat}(T)$ increases. Presumably, this occurs because the concentration of single spins in the samples grows, but these spins do not appear in the EPR. Such properties are characteristic of electrons in the conduction band. For these electrons, the spin relaxation time is short, so that there is no resonance absorption by these carriers. We cannot compare values of $\chi_{stat}(T)$ and $\chi_{dyn}(T)$ for this temperature range because of the difference between the spin concentrations manifested in each measurement technique.

In the liquid-helium and sub-liquid-helium temperatures range, both the static and dynamic magnetic susceptibilities increase as the temperature becomes lower. However, this increase occurs in different ways relative to the Curie law. The dynamic susceptibility increases more rapidly than according to this law, and the farther the spin concentration from the critical point, the faster this increase. At the same time, a wide scatter is observed in the growth rates of the static susceptibility. There are samples of three kinds: with faster rise, the same as that for the dynamic susceptibility; with a rise obeying the Curie law; and with a slower rise. In addition, a wide scatter of experimental values is observed in this temperature range. This is natural because the uncertainty in measurement results substantially grows when one large value is subtracted from another.

The numerical value of χ_{stat} at a temperature of 2 K fluctuates from 10^{-8} to 3×10^{-9} cm^3/g, but there is only any trend in the dependence of these values on the carrier concentration. Therefore, we can estimate the absolute value of χ_{dyn} with the small accuracy. Its variation at T = 2 K from 200 to 1400 relative units approximately corresponds to the scatter in χ_{stat}, but with a certain dependence on the carrier concentration observed for χ_{dyn}.

Comparison of Figs. 2 and 6 shows that the EPR technique can sufficiently precisely determine specific features of the interaction of spins at low temperatures, but the dependences for the static susceptibility, $\chi_{stat}(T)$, proved to be much less accurate. The large scatter of points, observed for the static susceptibility, masks true dependences. If these dependences are extrapolated with straight lines, they cross the abscissa axis in the vicinity of zero. This means that, to within the accuracy with which χ_{stat} could be measured, the samples are paramagnetic and no spin–spin interaction can be observed. At the same time, ESR measurements demonstrate that these interactions exist, but the accuracy with which the static susceptibility is measured is insufficient for their detection.

On the whole, comparison of the results obtained in measuring the paramagnetic susceptibility of semiconductors in the vicinity of the IM phase transition shows that the SQUID technique can also be used for this purpose, but it needs modification to improve its accuracy and sensitivity.

CONCLUSIONS

(1) Comparison of the results obtained by measuring the magnetic susceptibility by different methods shows that each of these has its own advantages and disadvantages: ESR gives the accurate temperature dependence of χ, but cannot determine its absolute value; SQUID gives the absolute value of χ, but has an insufficient accuracy for determining its temperature dependence.

(2) The temperature dependences of χ behave at low temperatures in qualitatively the same way, irrespective of the measurement technique. A decrease in temperature in the high-temperature range leads to a decrease in χ in SQUID measurements and to a weak increase in χ in ESR measurements.

ACKNOWLEDGMENTS

The study was financially supported by the Russian Foundation for Basic Research (grant no. 10-02-00629), Ministry of Education and Science of the Russian Federation (RF Presidential grant NSh-3306.2010.2), Presidium and Department of Physical Sciences of the Russian Academy of Sciences.

REFERENCES

1. N.F.Mott. Metal-Insulator Transitions 1974 (London: Tailor & Francis); 1979 (Moscow: Nauka)
2. Shklovsky B.I., Efros A.A *Electronic Properties of Doped semiconductors*, NY, Springer, 1984, Moscow: Nauka, 1979.
3. G.W.Ludwig and H.H.Woodbury. *Electron spin resonance in semiconductors.* Solid State Physics, edited by F.Seitz and D.Turnbull, v. **13**, NY, London Acad.Press, 1962.
4. K.Morigaki and S.Maekawa. J. *Phys. Soc. Jpn*. **32**, 462 (1972).
5. A.I. Veinger, A.G. Zabrodskii, T.V. Tisnek, S.I. Goloshchapov. *Semicond* **41**, 790, (2007).
6. A.I. Veinger, A.G. Zabrodskii, T.V. Tisnek, S.I. Goloshchapov. *Semicond* **42**,1274, (2008).
7. A.I. Veinger, A.G. Zabrodskii, T.V. Tisnek, S.I. Goloshchapov. *Appl. Magn. Res.* **35**, 439 (2009).
8. A.I. Veinger, A.G. Zabrodskii, T.V. Tisnek, S.I. Goloshchapov. *Semicond* **44,** 705, (2010).

Ageing and relaxation times in disordered insulators

T. Grenet[1], J. Delahaye[1] and M. C. Cheynet[2]

[1]Institut Néel, CNRS, BP 166, 38042 GRENOBLE cedex 9, France
[2]SIMAP, 1130 rue de la Piscine - BP 75 - F-38402 St Martin d'Hères cedex, France

thierry.grenet@grenoble.cnrs.fr

Abstract. We focus on the slow relaxations observed in the conductance of disordered insulators at low temperature (especially granular aluminum films). They manifest themselves as a temporal logarithmic decrease of the conductance after a quench from high temperatures and the concomitant appearance of a field effect anomaly centered on the gate voltage maintained. We are first interested in ageing effects, i.e. the age dependence of the dynamical properties of the system. We stress that the formation of a second field effect anomaly at a different gate voltage is not a "history free" logarithmic ($\ln t$) process, but departs from $\ln t$ in a way which encodes the system's age. The apparent relaxation time distribution extracted from the observed relaxations is thus not "constant" but evolves with time. We discuss what defines the age of the system and what external perturbation out of equilibrium does or does not rejuvenate it. We further discuss the problem of relaxation times and comment on the commonly used "two dip" experimental protocol aimed at extracting "characteristic times" for the glassy systems (granular aluminum, doped indium oxide...). We show that it is inoperable for systems like granular Al and probably highly doped InO_x where it provides a trivial value only determined by the experimental protocol. But in cases where *different* values are obtained like in lightly doped InO_x or some ultra thin metal films, potentially interesting information can be obtained, possibly about the "short time" dynamics of the different systems. Present ideas about the effect of doping on the glassiness of disordered insulators may also have to be reconsidered.

1. Introduction

Unexpected glassy features, involving slow relaxations of the electrical conductance, as well as electrical field effect anomalies in MOSFET samples, were discovered and studied in detail in insulating disordered indium oxide films [1] and then in granular [2] and ultra-thin films of metal [3]. They may be signatures of an electron glass state, where charge carriers are frozen by the effects of disorder and coulomb repulsions [4]. Two experimental salient features common to these systems are the very long relaxation times observed and the logarithmic with time ($\ln t$) dependence. More precisely, after a sample has been quenched from high T down to liquid He temperature, its electrical conductance is never stable and always decreases logarithmically with time. Correspondingly gate voltage (V_g) sweeps show the slow formation of a "dip" centered on the waiting V_g, which amplitude grows like $\ln t$. When V_g is rapidly changed (going out of the dip) the conductance increase indicates that the system is excited away from the equilibrium state towards which it was slowly relaxing, since the equilibrium state is believed to realize the minimal system conductance. If the new V_g value is maintained, the system starts a new relaxation towards a new equilibrium state corresponding to it, so that the first dip is progressively erased and a new one centered on the new V_g value, is formed. These phenomena are illustrated in the inset of Fig. 1a.

In this work we are interested in the dips dynamics. A pecularity of logarithmic relaxations is that no natural characteristic relaxation time can be defined, making it difficult to compare the dynamics of different samples or of a given sample in different conditions (e. g. at different temperatures). In section 2 we will show that at least in granular aluminium, the dynamics is more complex than first thought. In particular deviations from pure ln$t$ dip growth are observed and signal the occurrence of ageing effects, a hallmark of glasses. In section 3 we shall comment on a specific procedure which was designed to extract characteristic relaxation times. We will stress that this can be successful and reveal interesting phenomena only if deviations from the simple logarithmic behaviour are present. Our observations give incentives to study the short time parts (t less than a few seconds) of the relaxations, which haven't been investigated yet.

2. Ageing effects in granular Al thin films

2.1. "Simple" ageing protocol

Ageing is the fact that the dynamical properties of a non-equilibrium system evolve with time. Generally speaking, glasses become "stiffer" with age: the more they relax, the slower they respond to external stimuli. We have shown that ageing effects indeed exist in insulating granular Al thin films and first briefly recall here the basic results [5]. The growth of a *first* dip centered at V_{g1} after a quench from high temperatures is logarithmic with time, but it appears that the growth of a *second* dip (centered at V_{g2}) is not exactly logarithmic anymore. A deviation from pure ln$t$ is observed, which encodes the age of the system or the time t_{w1} during which the first dip was formed. This is illustrated in Fig. 1a where we show the growing amplitude of a second dip for different values of t_{w1}. At times spent at V_{g2} much smaller than t_{w1} the system is effectively old (its past history at V_{g1} is much longer than the time spent at V_{g2}) and the second dip growth is logarithmic, like in the lower curve of Fig. 1a for most of the time span. In the opposite limit (t_{w1} much smaller than the time spent at V_{g2}), the second dip growth is also logarithmic, consistently with the t_{w1} -> 0 limit for which one expects a history free pure logarithmic relaxation. But the two logarithms are not identical and at intermediate times of the order of t_{w1} the second dip growth accelerates to jump from one to the other. Thus the "s" shape observed for the middle curve in Fig. 1a. One may compute the difference between actual curves and pure ln$t$ (see Fig.1b) or smooth and differentiate the curves (see Fig. 1c). The resulting features are seen to depend on t_{w1} and, within our accuracy, to collapse on a single curve when plotted versus t / t_{w1} ("full ageing"). We note that the differentiated curves are similar to the ones observed in spin glasses [6]. In that case, the different spin contributions to the magnetization being additive, the curves are believed to represent the relaxation time distributions, which show an accumulation of relaxation times around $t = t_{w1}$.

It may be interesting to determine what kind of external perturbation does or doesn't erase the system's memory and thus rejuvenate it. Obviously an excursion to a high enough temperature does rejuvenate the system. We have shown [6] that a short excursion to a higher temperature at the end of t_{w1} just before switching the gate voltage from V_{g1} to V_{g2}, exponentially reduces the effective age of the system as determined from the growth of the second dip. This is consistent with the fact that a temperature increase has a strong effect on the dip: its width increases and its relative amplitude diminishes, and after the temperature is suddenly reduced back it takes some time for the dip to recover. This thermal erasure of the dip signals an erasure of the system's history memory and thus rejuvenates it.

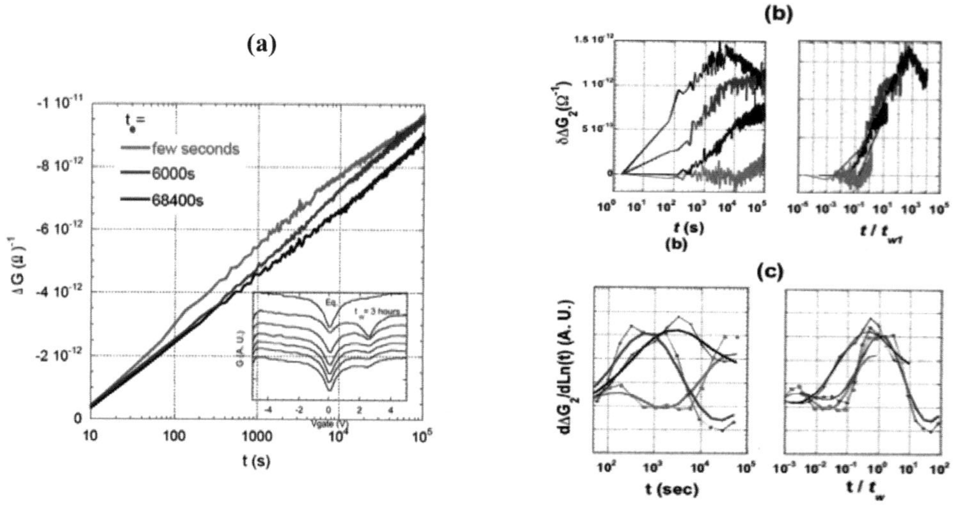

Fig. 1 Ageing effect in the growth of the second dip of a granular Al film (thickness: 20 nm, $R_\square$(4K) = 20 GΩ). (a): inset: typical conductance versus gate voltage curves after relaxation during t_{w1} under V_{g1}= 0 V (upper curve), after applying a positive second gate voltage V_{g2} for a time t_{w2} (second lower curve) and after setting back V_{g1}(subsequent curves). (a) main frame: second dip amplitude growth as a function of time for different t_{w1} values, from top to bottom curves: t_{w1} = 10 secs, 6000 secs and 6.85 10^4 secs; (b): departures from the pure Ln(t) growth as a function of t (left) and the reduced time t / t_{w1} (right); (c): derivative of ΔG_2 as a function of t (left) and the reduced time t / t_{w1} (right). The curves are shown for two filterings to show that the broad maximum position does not depend on the amount of filtering.

One may wonder what the effect of gate voltage changes is. Actually would any gate voltage change erase a system's memory, then ageing experiments would not be possible as switching from V_{g1} to V_{g2} would automatically rejuvenate the system. This is further illustrated by the following experiment: instead of keeping a constant V_{g1} during t_{w1} we apply a randomly changing gate voltage (distributed in the interval -5V / +5V). The result at t_{w1} is of course the absence of a well-defined dip (see Fig. 2a), but anyhow the system is "old" as shown by the subsequent growth of the second dip at V_{g2} which is characteristic of an "old" sample (see Fig. 2b). Thus although sudden changes of V_g excite the system to higher conductance states they do not immediately erase prior "correlations" established and so a history is memorized, even if no nice dip is allowed to form. One may conjecture that if the random V_g was changed fast enough so that the system does not have time to start relaxing at any V_g value then no history would be created.

To that respect thermal excitations are different, and qualitatively similar to applying high enough (i.e. non ohmic) bias voltages [7]. In this case during the application of the perturbation, the conductance is observed to relax slowly *upward*, showing that a slow excitation process goes on, whereas just after the application of a gate voltage change the conductance relaxes *downward*. It is possible that history erasure can be achieved by large enough gate voltage changes, as it was reported in lightly doped indium oxide films [8] but in granular aluminum this effect was not observed in the gate voltage range practically achievable.

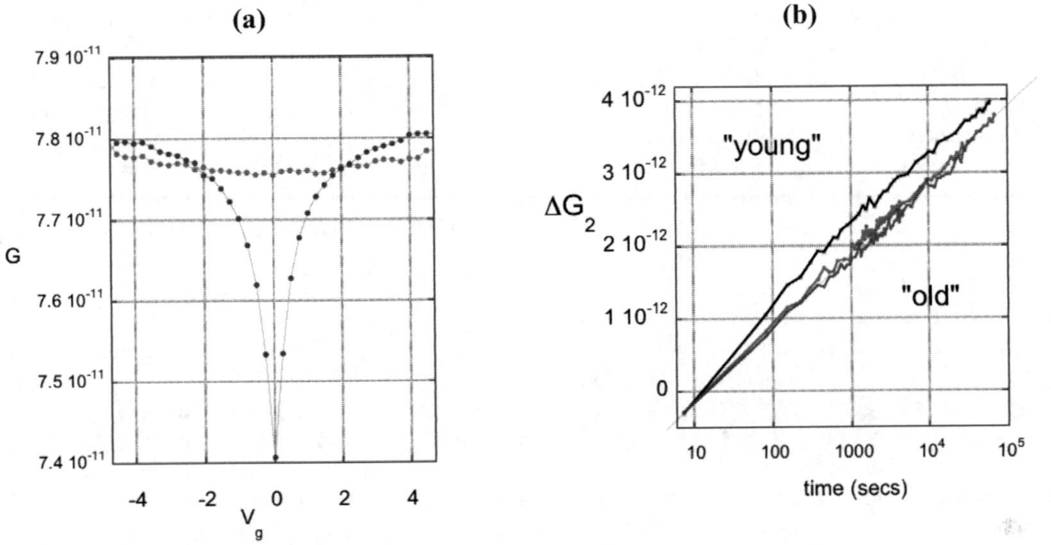

Fig. 2: (a): conductance versus gate voltage curves after a long relaxation under $V_{g1} = 0$ V (well formed dip) or after applying a gate voltage randomly varying in the interval [-5V,+5V] (almost flat curve); (b): second dip growth curves for the two cases of Fig.2 (a) (two lower curves). The two curves are almost undistinguishable and typical of an “old” system. They differ from the curve obtained for a “young” system shown for comparison.

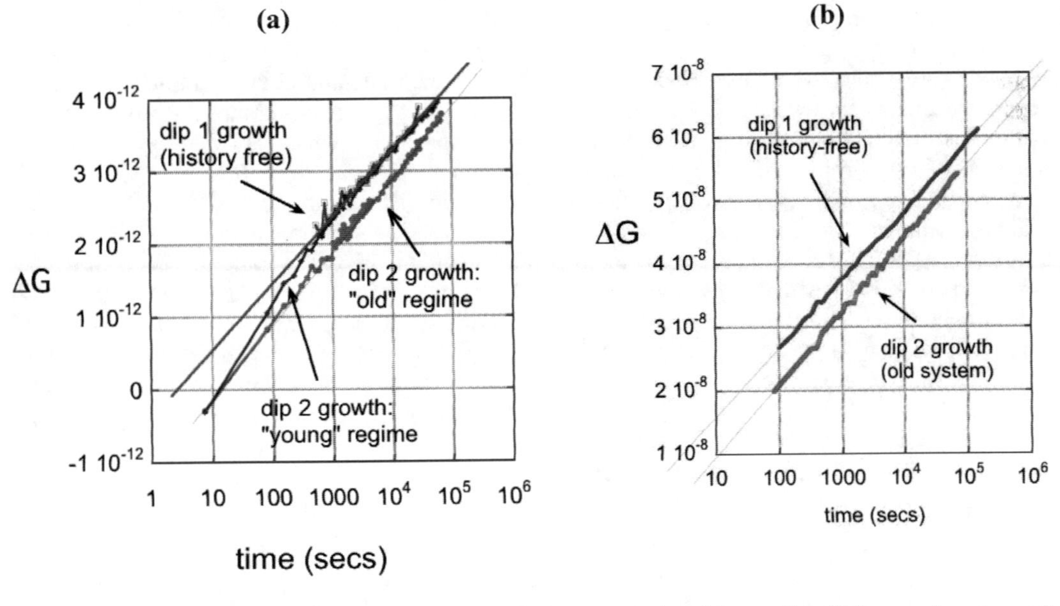

Fig. 3: (a): second dip growth curves in the “old” (lower curve) and “young” (middle curve) regimes. For t above 500 secs the *first* dip growth is also shown and is seen to be joined by the higher time part of the “young” curve. (b): first dip growth and second dip growth (in the “old” regime”) for another much less resistive sample. The difference in slope and vertical shift of the two logarithmic regimes is observed.

Journal of Physics: Conference Series **376** (2012) 012005 doi:10.1088/1742-6596/376/1/012005

We add a comment on the logarithmic regimes observed for the second dip growth in the "young" (time >> t_{w1}) and "old" (time<< t_{w1}) regimes. As noted above the two logarithms are different: the history free (i. e. first dip at V_{g1}) growth curve is above the other one but has a smaller slope than it. This feature is observed systematically, even in samples of very different insulating character as illustrated in Fig. 3. In the introduction we stated that generally speaking older systems respond slower than younger ones to external stimuli, but here the picture is not so clear: the logarithmic relaxation is on the contrary somewhat faster for the "old" case, but the "young" system is nevertheless more advanced because it started from a higher dip amplitude at the shortest times measured. Thus the younger systems are faster only at short times, which unfortunately cannot be explored using conductance measurements. One possible explanation is that the fast part of the history free relaxation occurs during the sample's quench profiting by higher relaxation rates at higher temperatures. The remaining unrelaxed "modes" then contribute a smaller logarithmic relaxation at longer times after the quench. However this scenario is at odd with known indications that the relaxations are not thermally activated [2], and the accelerated relaxation at times around t_{w1} is still to be explained. When extrapolated at longer times, the two logarithmic regimes cross for times of the order of a few tens of years. Clearly in this time range the system's behaviour must change, otherwise with t_{w1} values larger than this the switching from the "old" to "young" regime curves would imply a non monotonous second dip evolution, which seems unphysical. One may conjecture that the crossing time could be connected to the upper bound of the systems relaxation time distribution, but we presently see no simple way to check this hypothesis (apart from running a decades long experiment…).

2.2. "Two dip" protocol

In the electron glass litterature the term "ageing" was used with a different meaning [9] in the context of the so-called "two-dip" protocol. For clarity we briefly recall this. In these experiments one adds a further step in the protocol: after the gate voltage has been switched from V_{g1} to V_{g2} one waits for a duration t_{w2} and changes again the gate voltage. One is then interested in the second dip erasure. It is found that at short times the dip amplitude diminishes logarithmically with time and then of course starts to saturate when it approaches zero, for times of the order of t_{w2} (see Fig. 4). Thus the shape of the erasure curve is determined by t_{w2} and it was observed that when the experiment is repeated for different t_{w2} values, the curves are shifted in a logarithmic time scale and obey a simple scaling: they collapse on a single curve when plotted versus t / t_{w2} [9, 2]. This scaling was coined "ageing", which is probably misleading. Ageing manifest itself when one compares the dynamics of "young" and "old" systems, so we studied the erasure dynamics of the second cusp in the "old" ($t_{w1}>>t_{w2}$) and "young" ($t_{w1}<t_{w2}$) cases. We found that the collapsed curves are indeed different in the two limits: in the "old" case the logarithmic part of the curve extrapolates to zero at $t/t_{w2} = 1$, whereas in the "young" case it does so at $t/t_{w2} > 1$. This difference is the signature of ageing in the two-dip protocol. This is illustrated in Fig. 4 where it is seen that the extrapolation point varies continuously with t_{w1}. Thus the whole history is encoded in the erasure curves: the t / t_{w2} scaling encodes t_{w2} and the extrapolation value encodes t_{w1}. We also showed that the second dip erasure curves can be obtained from the relaxation curves of Fig1a: each erasure curve is the difference between a pair of relaxation curves shifted in time [5]. For "old" systems the erasure curve extrapolating to $t / t_{w2} = 1$ is the difference of two logarithms, in other cases erasure curves are the differences of two relaxation curves containing the departures from pure log discussed in the preceding section.

Although these ageing effects have not been studied in other systems like indium oxide, where only the "old" regime was investigated, there is no reason to think that they are not present.

Journal of Physics: Conference Series **376** (2012) 012005 doi:10.1088/1742-6596/376/1/012005

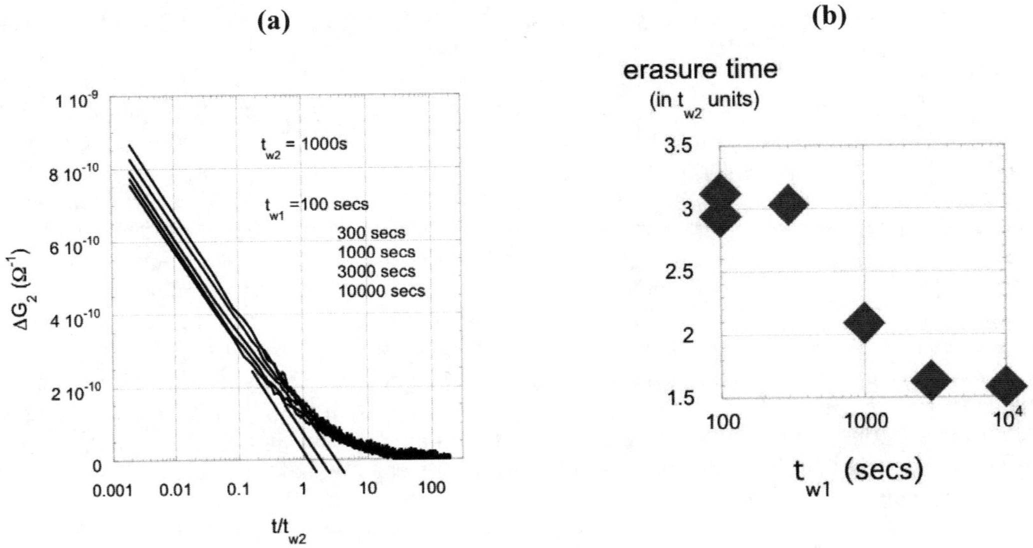

Fig. 4: (a): second dip erasure in the "two-dip" protocol. When plotted versus t/t_{w2} the curves do not coincide but also depend on t_{w1}, which determines at what time the extrapolated logarithmic parts cross the time abscissa. (b): "erasure time" (intercept with the time axis) as a function of t_{w1}.

3. Comments on the two-dip "relaxation times"

It was shown that in insulating indium oxide thin films electron doping has a clear effect on the dip width as well on its dynamics [10]. Understanding this phenomenon is crucial as it may explain why the very phenomena of interest here are not observed in "standard" doped insulators. In order to compare the dynamics of different samples (with different doping levels) a procedure was designed to determine a "characteristic relaxation" time. It amounts to let the system "equilibrate" for say a day or so after the quench under a gate voltage V_{g1}, then switch to V_{g2} and measure the time it takes for the curves of the two dips amplitudes to cross (the growing second dip becoming as large as the diminishing first dip, see sketch in Fig. 5). Would the first "equilibration" step (of duration t_{w1}) be long enough so that the system has indeed reached equilibrium, then the procedure would provide some characteristic relaxation time of the system. But if such is not the case the interpretation of the experiment may be quite different.

We have shown that in the case of granular aluminum films the time thus defined is not characteristic of the system and is essentially determined by the experimental protocol itself. Supposing, consistently with the known behaviour of the system, that the equilibrium state is not reached during the first step and that the first and second relaxations are essentially logarithmic with time with symmetrical slopes, one can show that the "relaxation time" measured is in fact:

$$\tau \approx \sqrt{t_{w1} t_{scan}} \quad (1)$$

where t_{scan} is the duration of one of the V_g scans repeatedly performed to measure the two dips amplitudes at different times. The square root dependence was checked for t_{w1} up to 70 hours (a value higher than the t_{w1} typically used) and for t_{scan} between 300 secs and more than 1100 secs [11].

In the case of indium oxide, it was shown that the "relaxation time" thus measured sharply increases with the electron doping level for $n_e < 10^{20}$ cm^{-3} and then saturates around 2 10^3 secs for higher carrier concentrations [10]. As it is known that highly doped indium oxide has logarithmic relaxations for times exceeding usual experimental times and as the 2 10^3 secs well corresponds to the

Journal of Physics: Conference Series **376** (2012) 012005 doi:10.1088/1742-6596/376/1/012005

estimate using equation (1) it is possible that the saturation of the "relaxation time" at high carrier concentration is actually an artifact and should not be given any physical significance. However of great interest are the cases where the determined "relaxation times" *differ* from the trivial value given by equation (1), as is the case in lightly doped indium oxide. This may mean that equilibrium was indeed reached during the first step, or that either the slopes of the logarithms are not symmetrical or the relaxations are not simply logarithmic. Inspection of the relaxation curves given in [12] show that both highly and lightly doped indium oxide samples have essentially logarithmic relaxations up to times exceeding the t_{w1} values. Moreover these logarithmic relaxations are slower for the supposedly fast samples (the lightly doped ones having small "relaxation times") than for the supposedly slow ones (the highly doped ones). Here again the difference seems to come from short time phenomena: in the lightly doped case the relaxations are already more advanced when the measurements start, especially the first dip erasure (see sketch in Fig. 5). It is still not clear whether this indicates a truly faster dynamics of the lightly doped samples, with a larger weight of the relaxation time distribution at small times, or if other effects like the fast erasure of memory induced by gate voltage changes, alluded to in our discussion of ageing and observed in lightly doped indium oxide samples, are important. Clearly further investigations of short time phenomena are needed in order to understand the effect of doping on the dynamics.

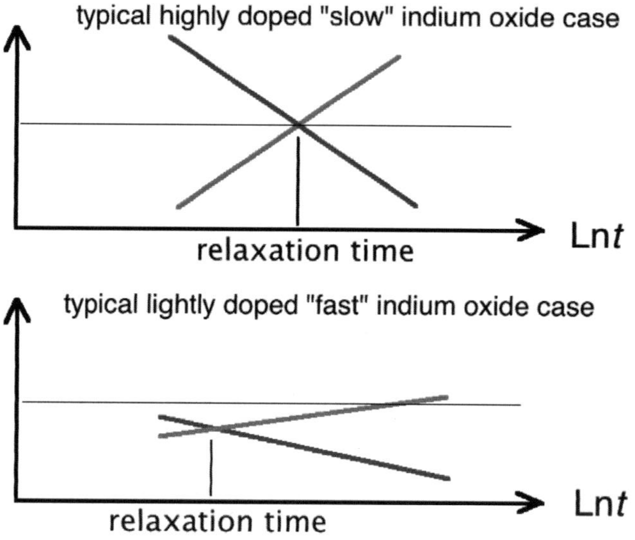

Fig. 5: Sketch of the relaxation curves used to determine the two-dip "relaxation time" in the case of high (upper sketch) and low (lower sketch) carrier concentration indium oxide insulating films (see original data in [12]).

4. Conclusion

We have been interested in the glassy behaviour observed at low temperature in the electrical conductance of some disordered insulators, attributed to an electron glass state. Although one salient feature of the systems involved is a nicely logarithmic relaxation after a quench from high temperatures, which can be observed over many decades in time, we have shown that richer phenomena and more physical information can be obtained in experiments where departures from the pure logarithmic relaxations arise.

One interesting example is the observation of physical ageing, characteristic of the glassy state. It manifests itself in specific departures from the pure logarithmic behaviour. These departures encode the "history" the systems experienced in their glassy state. Another example is related to the influence of carrier concentration on the dynamical properties of insulating indium oxide. In the lightly doped samples it seems that short time phenomena arise which make the dynamics faster than the logarithmic laws observed in the experiments at longer times. Further studies of this are needed, owing to the prime importance of the question of the role played by the carrier concentration in the appearance of the glassy state.

References

[1] BEN-CHORIN M., KOWAL D. and OVADYAHU Z., *Phys. Rev. B*, **44** 3420 (1991)
BEN-CHORIN M., OVADYAHU Z. and POLLAK M., *Phys. Rev. B*, **48**, 15025 (1993)
[2] GRENET T., *Eur. Phys. J. B*, **32**, 275 (2003)
GRENET T. *et al.*, *Eur. Phys. J. B*, **56** 183 (2007)
[3] MARTINEZ-ARIZALA G. *et al.*, *Phys. Rev. Lett.*, **78** 1130 (1997)
See contribution by A. FRYDMAN and co-workers in these proceedings
[4] DAVIES J. H., LEE P. A. and RICE T., *Phys. Rev. Lett.,* **49** 758 (1982)
M. GRUNEWALD et al., J. PHYS. C **15**, L1153 (1982)
M. POLLAK and M. ORTUNO, Sol. Energy. Mater. **8**, 81 (1982)
[5] GRENET T. and DELAHAYE J., Eur. Phys. J. B **76**, 229–236 (2010)
[6] VINCENT E. in *Ageing and the Glass Transition (Lecture Notes in Physics)* edited by HENKEL M., PLEIMLING M. and SANCTUARY R. (Springer, 2007)
[7] V. ORLYANCHIK and Z. OVADYAHU, Phys. Rev. Lett. **92**, 066801-1 (2004)
[8] OVADYAHU Z. and POLLAK M., *Phys. Rev. B*, **68** 184204 (2003)
[9] A. VAKNIN, Z. OVADYAHU and M. POLLAK, Phys. Rev. Lett. **84**, 3402 (2000)
[10] A. VAKNIN, Z. OVADYAHU *and* M. POLLAK, Phys. Rev. Lett. **81**, 669 (1998)
[11] T. GRENET and J. DELAHAYE, arXiv:1105.0984
See also: A. AMIR, Y. OREG and J. IMRY, *Annual Review of Condensed Matter Physics*, **2,** 235 (2011)
[12] VAKNIN A., OVADYAHU Z. and POLLAK M., *Phys. Rev. B* **61**, 6692 (2000)

Quantum Coulomb gap

M. Pino,A. M. Somoza and M. Ortuño
Departamento de Física - CIOyN, Universidad de Murcia, Spain

Abstract. We study numerically strongly-localized systems with long-range interactions. We obtain the density of states for 1-D Anderson Hamiltonian with $r^{-1/2}$ interactions by exact diagonalization. The limit in which the hopping term is zero has been well studied and leads to a linear gap near the Fermi energy. We find the same linear dependency for non zero hopping however the slope decreases. The effects of possible numerical problems, like finite size effects or the use of a limited set of eigenvalues, is analyzed.

1. Introduction

From the work of Mott [1] dealing with the transport properties of disorder system a vast field of solid state physics has merged. That field is known as Variable Range Hopping (VRH) and its main concern is the characterization of the conduction properties of disorder interacting systems at finite temperature.

A fundamental step in that task is to understand the density of states (DOS) of disorder and interacting system. In 1971 Pollak [2] suggested the existence of a gap in the DOS when Coulomb like interactions are present in disordered and completely localized electronic systems. This phenomenon is better known as the Coulomb gap and should produce dramatic consequences in the low temperature conduction. The existence of that gap has been experimentally observed in [3] and [4].

Although the theory in which the Coulomb gap appears has been already established by Efros and Shklovskii [5, 6], it has been always assumed that the systems under study are well represented by classical laws. At a first stage, it could be thought that in the strongly-localized regime it is possible to renormalize the localization length of electrons and obtain that they are completely localized. Nevertheless it is clear that quantum effects must play a role at least near the metal-insulator transition. Quantum effects in the coulomb gap has been poorly studied [7, 8, 9, 10], just because it is a long-range effect and within an exact quantum treatment it is impossible to deal with large systems. In these studies different approximations were used and obtained opposite results, in some cases the authors claimed that the Coulomb gap opens due to quantum effects and in others either remains the same or closes.

In this work we study the generalization of the Coulomb gap when quantum effects are taken into account. The DOS near the Fermi energy for a 1-d lattice with up to 20 sites is calculated. As a Coulomb interaction ($1/r$) produces a logarithmic Coulomb gap when dealing with 1-D system, we use a $\frac{1}{\sqrt{r}}$ interaction which gives a linear gap similar to the usual Coulomb gap in two dimensions.

2. The model

We consider a one-dimensional lattice system with the following Hamiltonian:

$$H = \sum_i \phi_i n_i - t \sum_{<i,j>} c_i^\dagger c_j + V \sum_{i \neq j} \frac{(n_i - K)(n_j - K)}{\sqrt{r_{ij}}}, \tag{1}$$

where ϕ_i is a random site energy, with a uniform distribution in the interval $[-W/2, W/2]$, t is the hopping parameter, V is the strength of interaction and K is the compensation constant; $< i, j >$ indicates a sum over all nearest neighbors pairs. As mentioned above, we have chosen an interaction potential decaying as $1/\sqrt{r}$ in order to obtain a classical linear gap. In the following we fix the disorder amplitude to $W = 2$ and the interaction strength to $V = 1$.

The main goal of this work is to study the single-particle density of states (DOS) for a given number of electrons N and size of the system L. We split this DOS into two contributions, one for holes, extracting an electron out of the system, and one for electrons, adding an electron to the system. The local DOS for holes at site k, which by assumption have negative energies, is

$$\rho_h^N(\epsilon, k) = \sum_\alpha \delta\left(\epsilon + E_\alpha^{N-1} - E_0^N\right) \left|\langle\psi_\alpha^{N-1}|c_k|\psi_0^N\rangle\right|^2, \tag{2}$$

while the DOS for particles is

$$\rho_p^N(\epsilon, k) = \sum_\alpha \delta\left(\epsilon - E_\alpha^{N+1} + E_0^N\right) \left|\langle\psi_\alpha^{N+1}|c_k^\dagger|\psi_0^N\rangle\right|^2, \tag{3}$$

where the superindex in the energy refers to the number of electrons in the sample and the subindex to the energy level (zero referring to the ground state). We shall employ the grand canonical ensemble, search for the ground state of the hamiltonian for different number of electrons and keep the number N that minimizes the grand potential $\Omega = H - \mu N$. In the classical limit this is a rather expensive procedure compared to other procedures (see for example [11]), but as we will see it is not so expensive in the quantum case and in this way we ensure that the electron energies are always positive (and the hole energies negative).

In order to compute the density of states, Eqs. 3 and 2, we also need to partially diagonalize the problem with both $N + 1$ and $N - 1$ electrons obtaining only a set of the smallest eigenvalues and their eigenvectors in each case. Then, Eqs.3 and 2 are computed and the results stored in an histogram.

As we are interested only in the DOS near the Fermi level we need to get a small number, ns, of low energy states with one more and one less electron than in the ground state. This part is the most expensive. Nevertheless we do not need to perform full diagonalization of the hamiltonian which would be much more expensive. All the numerical data were obtained using an eigensolver based on ARPACK libraries. Those libraries are specially suited to work with sparse matrices and to compute a few eigenvectors rather than fully diagonalize the Hamiltonian. We use system sizes from $L = 10$ to $L = 20$ and for the disorder average we compute over 30000 samples.

As an example we show in FIG 1 the classical DOS ($t = 0$) using this procedure for different values of ns. We see that all curves overlap near the center of the gap and for $ns > 20$ the gap is properly represented. For this system size, with $ns > 60$ we already reproduce the full single-particle DOS. Another more rigorous procedure to establish a range of energies in which our computation is valid is the following. For a given ns, we set up a maximum energy $\epsilon_{max}(ns)$ in such a way that all the samples used in the computation obey $E_{ns} > \epsilon_{max}(ns)$. Then, the error due to the truncation in the range $[-\epsilon_{max}, \epsilon_{max}]$ will be exactly zero. We found that this condition is too strong. Instead, we adopt ϵ_{max} in such a way that only 1% of the samples violate the inequality $E_{ns} > \epsilon_{max}$. This error is small enough to explore the relevant physics of the problem. The different values of ϵ_{max} are shown in the figure as vertical dashed lines. For all calculations of the DOS we have performed the same analysis, similar vertical lines will be plotted in other figures in this work. Another way to show the effect due to the truncation in the number of eigenvectors used it to check the area, A, of the numerical DOS. We observe in the legend of FIG 1 that, for this system size, $ns = 60$ is enough to provide accurate results in the whole energy range.

We may notice in FIG 1 that the DOS presents some filling of the gap near zero, this is a well-known finite size effect. In the classical case, the DOS near the Fermi energy is completely dominated by long

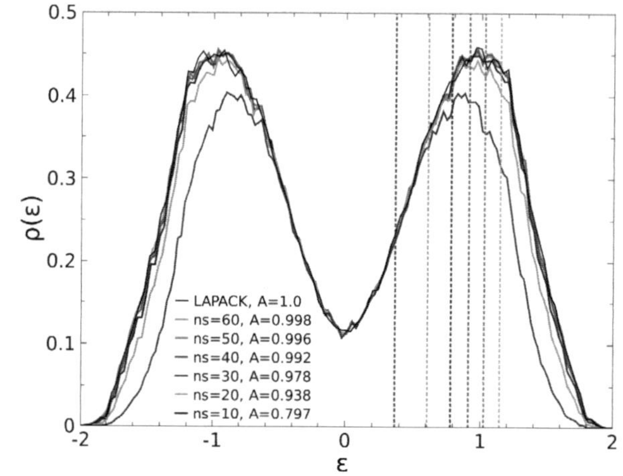

Figure 1. The classical single-particle DOS obtained retaining only a number of states ns. The size of the system is $L = 10$ and $K = 1/2$. Different colors represent different values of ns. The black solid line has been obtained using the LAPACK algorithm which calculates all the spectrum and thus, this is an exact computation. In the legend is specified the area, A, of each curve. See text for further explanations.

distance correlations, so for small systems this part of the DOS cannot be properly obtained. As the sizes available with our method are rather small we need a careful analysis of the effect of increasing L. In Fig. 2 different system sizes are used to compute the classical Coulomb gap. Though the finite size effects are important at a region close to the Fermi energy we can observe that for an intermediate part of the gap they are quite small. This is the region where we will concentrate our analysis. Again we may notice the vertical lines which correspond to ϵ_{max} for each case, we see that the region of interest is well inside the interval $[-\epsilon_{max}, \epsilon_{max}]$.

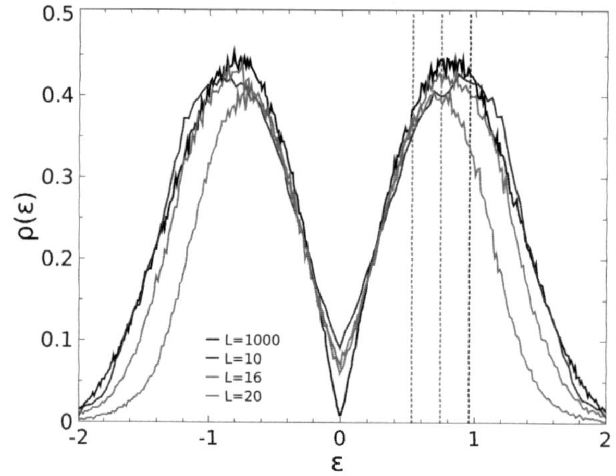

Figure 2. Several system sizes are depicted for the classical DOS. The dashed vertical lines indicate again the accurate energy range as described in the text related to FIG 1. The case $L = 1000$ has been computed using a classical algorithm, which allows for much bigger system sizes.

Related to FIG. 2, we should mention that, according to the self-consistent theory [6] the slope of the DOS should be $1/2$. The intermediate part of the gap (excluding the region near zero) fits very well to a straight line with slope 0.8. This subject deserves more attention, but it would require to study much larger systems, so here we will concentrate in the quantum effects.

3. Quantum Effects

Once we have clarified the role of the two main sources of error we introduce a small hopping t in order to estimate the quantum effects. The results are shown in FIG. 3. It clearly shows that in a strongly localized regime, quantum effects tend to open the gap. The gap remains fairly linear but with a smaller slope. This is the main result of this work. This result is in contradiction with previous works

FIG. 3 is not affected by the truncation in the number of eigenvalues used. In fact only the region $(-\epsilon_{max}, \epsilon_{max})$ is plotted for each case, but it is important to verify that it is not affected by finite size effects either. In FIG. 4 we show the results for $t = 0.2$ for different system sizes. We observe that although the finite size effects are important at a region close to the Fermi energy, for a intermediate part of the gap they are quite small. In fact the effects seem to be smaller that in the classical case, FIG. 2.

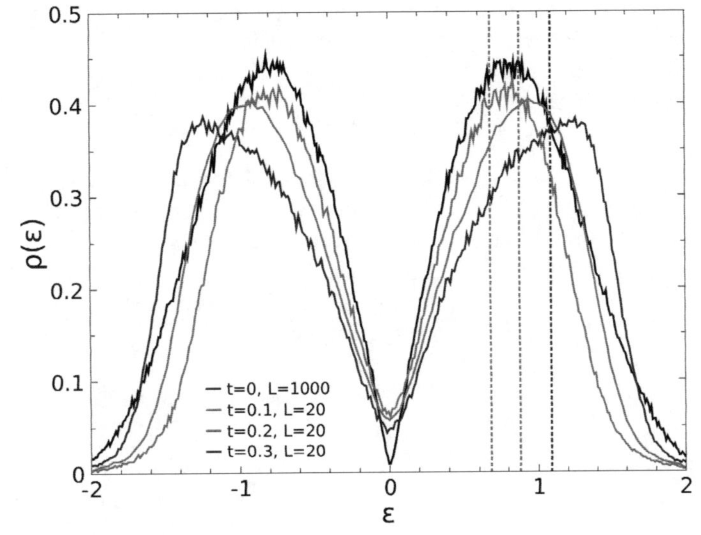

Figure 3. Single-particle DOS versus energy $L = 20$ and several values of t. The black line corresponds to the classical DOS for a large system size.

One possible explanation for the opening of the gap, could be the tendency to form a periodic structure. Without disorder, the ground state would correspond to a kind of anti-ferromagnetic order, with charges of opposite sign placed in nearest neighbors. Obviously, in the classical limit we do not get such long-range order. But, it is true that the disorder selected is not very large, our system sizes are small, so a correlation length of the order of the system size would be seen as a long-range order. Also, the hopping term could partially favor this anti-ferromagnetic order as an electron with the neighbors empty can delocalize and reduce its energy; similarly to the anti-ferromagnetic order in the Hubbard model [12]. So it is necessary to check that this is not the situation in FIG. 3. We have calculated the correlation function

$$C_{i,j} = \langle \psi_{gs} | n_i (1 - n_j) | \psi_{gs} \rangle, \tag{4}$$

where ψ_{gs} is the ground state and $n_i = C_i^+ C_i$. We have checked that certainly there is short-range correlation but it tends to zero as sites i and j separate apart. Another way to check that this effect is

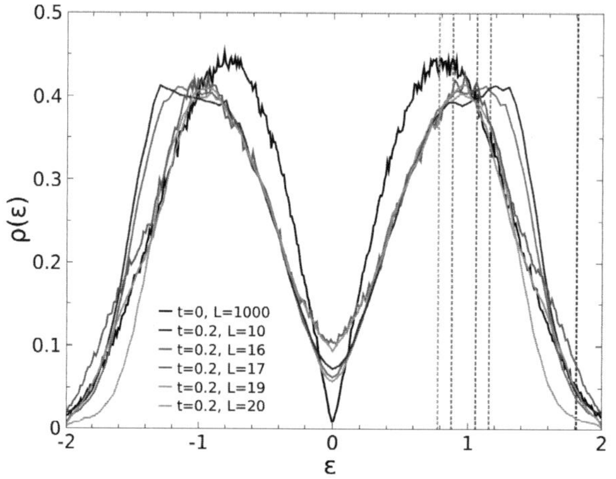

Figure 4. Single-particle DOS versus energy for several system sizes and $t = 0.2$. The dashed vertical line indicates again the accurate energy range as described below. The case $L = 1000$ corresponds to the classical limit and is shown only for comparison.

not affecting our results is to study systems with an odd number of sites (see FIG. 5). We can see that, certainly, in the intermediate part of the gap the system is not affected by any kind of finite size effect. In the center of the gap there are differences between odd and even system sizes, the later slightly favor correlations and produce a slightly deeper gap. Although, in the center the tendency indicates that the depth decreases as we increase system size, due to the large finite size effect, it is not possible to conclude anything concrete about the shape of the gap very near the Fermi level, although the DOS seems to go to zero at E_{F}.

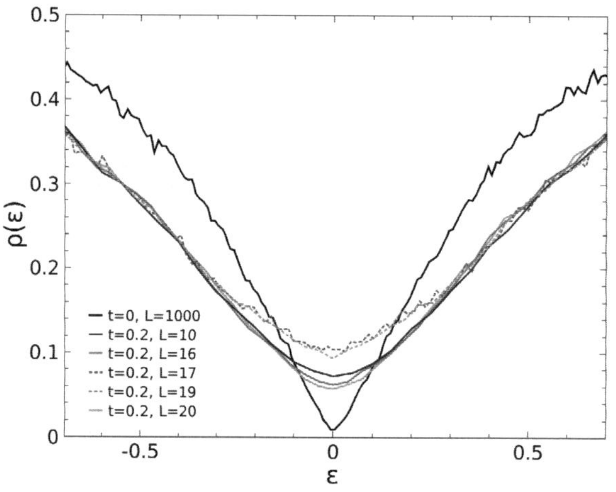

Figure 5. Single-particle DOS versus energy for several system sizes and $t = 0.2$ An odd-even effect can be appreciated near the Fermi level, this effect disapears for intermediate energies inside the gap. The case $L = 1000$ corresponds to the classical limit and is shown only for comparison.

4. Estimation of the localization length

In order to better understand the results of the previous sections it is necessary to estimate the localization length, ξ, of the system for the range of t used. In order to extract this length we have used two equivalent methods. On one hand, we can compute the sensibility to a change in the boundary conditions. Let us denote $E(\phi)$ the energy of a system under boundary conditions $c_{L+1} = e^{i\phi} c_1$, and $\Delta E = E(\pi) - E(0)$ the difference between periodic and anti-periodic boundaries. In a similar way as in [13] [14], the localization length could be extracted from:

$$\lim_{L\to\infty} \log\left(\Delta E(-1)^N\right) = -\frac{L}{\xi} + b, \tag{5}$$

where N is the number of electrons in the sample and b a constant. On the other hand a two points correlator could be used. In this case, the localization length is given as in [15] by:

$$\lim_{L\to\infty} \log\langle\psi_{gs}|c_L c_1^\dagger|\psi_{gs}\rangle = -\frac{L}{\xi} + b'. \tag{6}$$

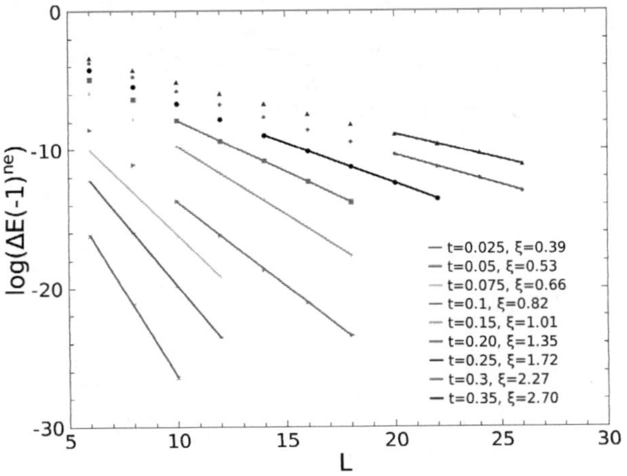

Figure 6. Average of the $\log \Delta E$ versus system size, L, for several values of t. In the legend we can see the localization length for each case. The straight lines show the fit to eq. 5 for the final points in each case.

We have checked that both methods lead to equivalent results. We have found that with the method of twisting the boundary conditions we get slightly better numerical precision. We performed two diagonalizations to compute the ground state energy for periodic and anti-periodic boundary conditions and average over 500 samples. In Fig. 6 the average of the $\log \Delta E$ is represented and from there the localization length is computed.

We see that in some cases, for our system sizes we do not get a perfect straight line, Nevertheless, for $t = 0.1$ and $t = 0.2$ we can estimate the localization length by roughly 0.8 and 1.0 respectively, of the order of the nearest neighbors distance.

5. Conclusions

We have studied numerically a quantum strongly-localized systems with long-range interactions. We have calculated the single-particle DOS for systems with localization length of the order of nearest neighbors distance.

Journal of Physics: Conference Series **376** (2012) 012006 doi:10.1088/1742-6596/376/1/012006

In our case, the limit in which the hopping term is zero has been well studied and leads to a linear gap near the Fermi energy. We find the same linear dependency for non zero hopping however the slope decreases. We have analyzed the possible numerical problems, like finite size effects or the use of a limited set of eigenvalues, showing that the behavior in the intermediate part of the gap is not affected by such problems.

We would like to acknowledge support from DGI Grant No. FIS2009-13483.

6. Bibliography

[1] N. F. Mott. Phil. Mag. 19:835 (1969)
[2] M. Pollak. Discuss. Faraday Soc. 50 (1970)
[3] J. G. Massey and M. Lee, Phys. Rev. Lett. 75, 4266 (1995)
[4] V. Y. Butko, J. F. Ditusa, and P. W. Adams, Phys. Rev. Lett. 84, 1543 (2000)
[5] A. L. Efros and B I Shklovskii, J. Phys. C : Solid State Phys. 8, L49 (1975)
[6] A. L. Efros, J. Phys. C: Solid State Phys. 9, 2021 (1976)
[7] G. Vignale and W. Hanke, Phys. Rev B 34, 3003 (1986)
[8] Q. Li and P. Phillips, Phys. Rev B 48, 15035 (1993)
[9] F. Epperlein, M. Schreiber and T. Vojta, Phys. Rev B 56, 5890 (1997)
[10] G. Sang Jeon, H-W. Lee and M. Y. Choi, Phys. Rev B 59, 3033 (1999)
[11] S. D. Baranovskii, A. L. Efros, B. L. Gelmont and B I Shklovskii, J. Phys. C: Solid State Phys. 12, 1023 (1979)
[12] G. D. Mahan, *Many-Particle Physics*, page 416, Kluber, New York (2000).
[13] P. Schimtteckert, T. Schulze, C. Shuster, P. Schwab and U. Eckern, Phys. Rev. Lett. 80, 560 563 (1998)
[14] P. Schmitteckert, R. A. Jalabert, D. Weinmann, and J.L. Pichard, Phys. Rev. Lett. 81, 2308 2311 (1998)
[15] J. M. Carter and A. Mackinnon, Phys. Rev. B 72, 024208 (2005)

Journal of Physics: Conference Series **376** (2012) 012007 doi:10.1088/1742-6596/376/1/012007

Numerical studies of relaxation in Electron Glasses

M. Ortuño and A. M. Somoza
Departamento de Física - CIOyN, Universidad de Murcia, Spain
E-mail: moo@um.es

Abstract. We study numerically strongly localized systems with long-range Coulomb interactions. We obtain a set of low-energy many-electron configurations. We calculate transition rates between states involving up to six electron jumps. We obtain the density of configurations and the density of metastable states. We also analyze the stability of these metastable states. We use the master equation to simulate the dynamic of the system, both without and with an applied electric field. We obtain in this way the energy relaxation from excited states and the behaviour of the conductivity as the system relaxes. We observe a drastic change in this behaviour for relaxation times close to the Maxwell time.

1. Introduction

Coulomb glasses are systems with localized electronic states and Coulomb interactions between carriers. Very slow relaxation rates are commonly observed in these systems due to the exponential dependence of the transition rates with hopping length and energy and due to the many-valley structure of the phase space.

Experimental studies on indium oxides by the group of Ovadyahu [?], have shown typical glassy phenomena such as logarithmic relaxation and aging. Similar effects have been found on granular metals [2], on ultrathin metal films [3] and on ferromagnetic granular nickel (Frydman).

Slow relaxation in Coulomb glasses has been studied theoretically (Pollak, Yu, Weizmann, Müller) and by Monte Carlo (MC) simulations (Grempel, Efros, Murcia).

2. Model and numerical procedure

We consider two–dimensional Coulomb glasses in the strongly localized regime, described by the standard tight–binding Coulomb gap (CG) Hamiltonian [?, ?]

$$H = \sum_i \epsilon_i n_i + \sum_{i<j} n_i n_j \frac{1}{r_{ij}} \ , \tag{1}$$

where $n_i = 0, 1$ are occupation numbers and ϵ_i are the random site energies chosen from a box distribution with interval $[-W/2, W/2]$. r_{ij} is the distance between sites i and j. We consider square samples of lateral size L and sites placed at random. $l_0 = L/\sqrt{N}$ is our unit of distance and $1/l_0$ is our unit of energy and temperature. We study systems ranging from 500 to 2000 sites, and establish a minimum separation between sites equal to 0.2. The range of the disorder is $W = 2$ and the localization length a is equal to unity. We choose the number of electrons to be half the number of sites, and use cyclic boundary conditions.

We used a numerical algorithm to obtain the ground state and 2.5×10^5 lowest energy many–particle configurations of the system. The algorithm [?, ?] consists of the following two stages. In the first stage, we repeatedly start from states chosen at random and relax each sample by means of a local search procedure. In an iterative process, we look for neighbors of lower energies and always accept the first such state found. The procedure stops when no lower energy neighboring states exist, which insures stability with respect to all one–electron jumps and compact two–electron jumps. We then consider a set of metastable states found by the process just described and look for the sites which present the same occupation in all of them. These sites are assumed to be frozen, *i.e.* they are not allowed to change occupation, and the relaxation algorithm is now applied to the unfrozen sites. The whole procedure is repeated until no new frozen sites are found with the set of metastable states considered.

In the second stage, we complete the set of low–energy configurations by generating all the states that differ by one– or two–electron transitions from any configuration stored. In order to speed up this process, which is very CPU time consuming, we again assume frozen and unfrozen sites and in first place look for neighboring configurations by changing the occupation of unfrozen sites only. We later relax this restriction in the final stage.

We are also interested in storing metastable states and we have developed a variant of the previous algorithm to this aim. In this case, we only relax by one-electron hops of length smaller than a critical value that we take equal to 6. This allows us to obtain metastable states with respect to jumps shorter than the critical one and also with respect to one-electron transitions. In this version of the algorithm we do not employed the second stage of the original method and we consider a huge number of initial random configurations.

To check the degree of completeness of the set of metastable states obtained, we have plotted in figure the evolution of the energy of the highest energy configuration stored as a function of the number of relaxation processes attempted on a logarithmic scale. Each relaxation process consists in choosing an initial random configuration and let it relax by one-electron hops until a metastable state is reached. Each of the four curves in figure corresponds to a different disorder realization, i.e. position of the sites and random site energies. The curves start when 250.000 different metastable states have been stored. We note that the convergence is fairly good when the number of attempts is larger than $\approx 10^8$. There are important fluctuations from sample to sample in the starting point of each curve and in the final energy, reflecting the variations in the complexity of the different energy landscapes.

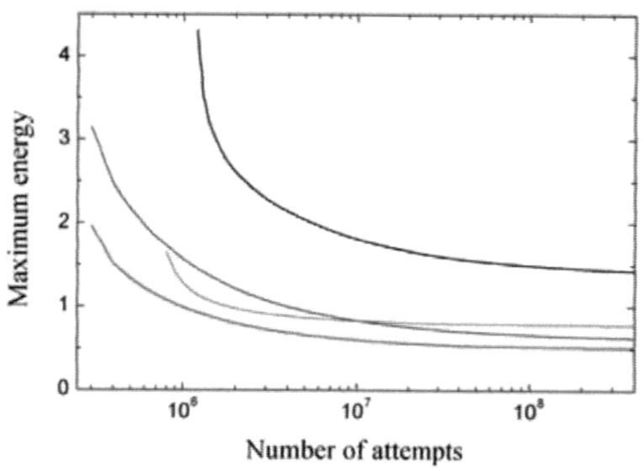

Figure 1. Energy of the highest configuration stored as a function of the number of relaxation attempts.

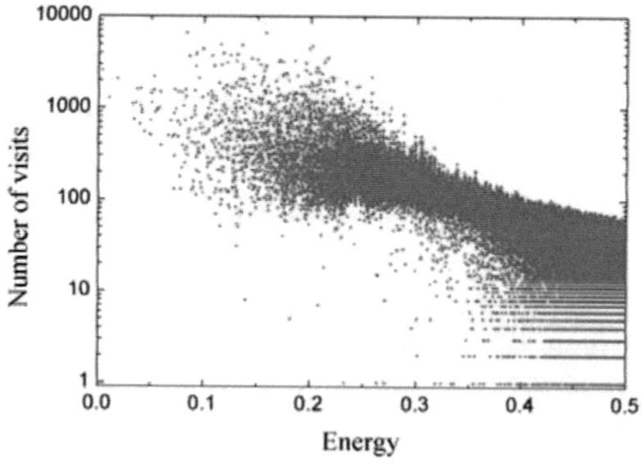

Figure 2. Number of times that a configuration is reached from different initial configurations as a function of the energy of the configuration.

Journal of Physics: Conference Series **376** (2012) 012007 doi:10.1088/1742-6596/376/1/012007

In figure 2 we have plotted the number of times that a configuration is reached from any initial random state versus the energy of the final configuration. We notice that for energies in the first half of the figure each configuration is visited at least almost ten times and often more than a thousand times. It is unlikely that there are configurations in this energy range that have not been visited at all. For energies above 0.4 (for the disorder realization of figure 2 there are many states that have been reached once, and we expect to be also many other states that still have not been visited at all.

3. Master equation

We have studied the time evolution of our system through the master equation in configuration space. The system is described by the time dependent probability vector, $\mathbf{p}(t)$, whose components are the occupation probabilities of the different many-electron configurations. The evolution depends on the transition rates between configurations. In this work, we assume that the system evolves through single–electron hops. The transition rate for a jump from site i to site j when the system is in configuration α is

$$\Gamma_{ij}^{(\alpha)} = \tau_0{}^{-1} \exp\left(\frac{-2r_{ij}}{a}\right) \min\left[1, \exp\left(-\frac{E_{ij}^{(\alpha)}}{kT}\right)\right] \quad (2)$$

where τ_0 is the inverse phonon frequency, of the order of 10^{-13} s.

The master equation setting the evolution of our system is

$$\frac{d\mathbf{p}(t)}{dt} = \mathcal{A}\mathbf{p}(t) \quad (3)$$

The dynamical matrix $\mathcal{A}$ gives us the rate of change from one configuration α to another β. Its non–diagonal matrix elements are $\Gamma_{ij}^{(\alpha)}$, if configurations α and β differ by a single jump from i to j, and zero otherwise. Its diagonal elements are $-\sum_{i,j} \Gamma_{ij}^{(\alpha)}$, so that the total probability is conserved.

We choose for the initial state the mixture of all configurations with weights equal to the Boltzmann factor for a temperature 40 times larger than the phonon bath temperature, which in this study we take as $T = 0.002$. We checked that the results do not depend on the initial state considered, provided it is widely distributed between configurations.

When we want to study how it relaxes the conductivity of the system, we simulate the application of an electric field F with an extra energy contribution equal to $-FSum_i x_i n_i$, where x_i is the coordinate along the direction of the field of site i. The inclusion of this term modifies the transition rates, equation (2), depending on whether the transition is in favour of or against the electric field.

In order to characterize the degree of stability of the metastable states, we have considered an extension of equation (2) for the transition rates of many-electron jumps. We have assumed that in this case equation (2) is still valid provided that r_{ij} is substituted by the total hopping distance of all electrons involved and E_{ij} by the energy difference between the final and the initial configurations. The total hopping distance is minimized with respect to all possible transitions between the initial and final occupations.

4. Density of metastable states

We have first calculated the density of configurations as a function of energy. In figure 3 we represent the integrated density of many-electron configurations as a function of energy (black, upper curve). We have also obtained the densities of metastable states according to their degree of stability and their accumulated values are also plotted in figure 3. The red curve corresponds to configurations that are stable with respect to all one-electron transitions, i.e. that cannot lower its energy via a single electron jump, but there is a two electron transition that do lower its energy. The green and blue curves correspond to accumulated densities of states stables with respect to two and three electron transitions, respectively. Finally the magenta dashed curve represents the accumulated density of configurations that can lower their energy by a single electron jump longer than the critical hopping distance considered, 6 in our case.

All curves in figure 3 can be fitted relatively well by expressions of the form

$$\log N(E) = A \exp\{BE^{1/2}\} \tag{4}$$

with A and B constants. It is surprising that there is no sign of saturation of the density of metastable states up to the relatively high energies considered. Also, except for a small crosover at low energies between the curves for metastable states with respect to one- and two-electron jumps, there is no predominance of highly stable states at low energies, as could be reasonably expected.

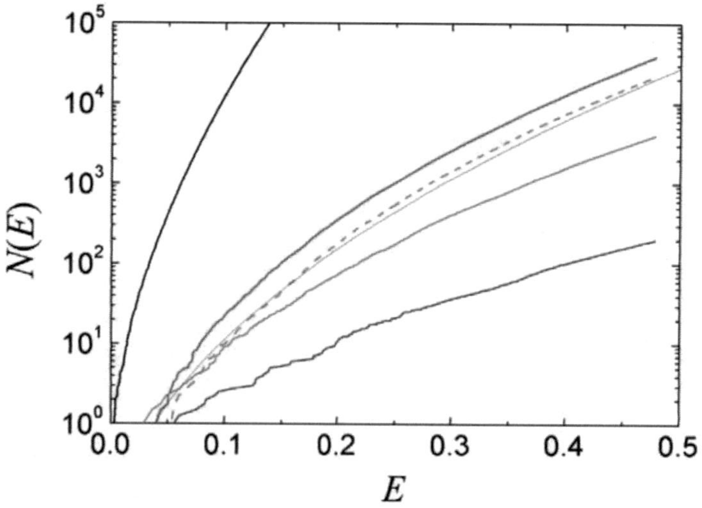

Figure 3. Integrated density of configurations as a function of energy for different degrees of stability. The black curve corresponds to all configurations, the red, green and blue curves to configurations stable with respect to one, two and three electron jumps, respectively. The magenta dashed curve corresponds to states stables with respect to one electron jumps shorter than the critical hopping distance.

5. Stability of metastable states

For each of the metasable configurations found, we calculate all transitions downward in energy to other metastable states. We consider $T = 0$ and so we concentrate only in the total hopping distance. It is clear that the direct transition between two metastable states, i.e. the bottom of two valleys, is not in general the most effective transition between the valleys, but we assume that represents a rough estimate of the stability of the valleys. For each metastable state we store the fastest of all previously calculated transitions downward in energy. In figure 4 we have plotted the hopping distances of the fastest relaxation transition from any metastable state arranged in decreasing order. These distances are proportional to the logarithm of the transition

Journal of Physics: Conference Series **376** (2012) 012007 doi:10.1088/1742-6596/376/1/012007

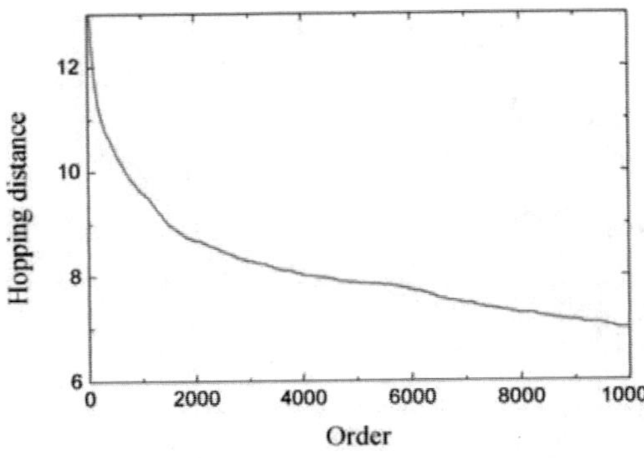

Figure 4. Hopping distance of the optimal transition from metastable states arranged in decreasing order.

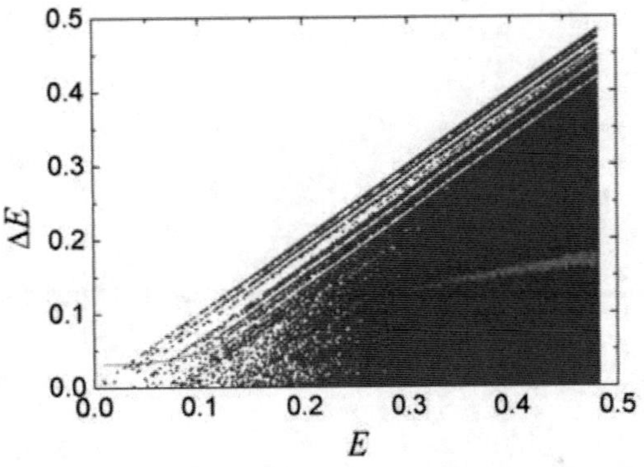

Figure 5. Energy relaxed in the optimal transition from metasble states versus the initial energy.

rates and so to the logarithm of the relaxation times. A constant distribution of these logarithms should corresponds to a straight line in our figure.

In figure 5 we have represented the energy decrease of the fastest relaxation transitions as a function of the energy of the relaxing configuration. When a transition ends up in a configuration with a larger transition rate (shorter relaxation time) than the original one we add to the energy difference of the transition the energy decrease of the subsequent transition. The red line in figure 5 is a smoothing of the data. We see that on average the energy difference of the transitions grows linearly with the initial energy, although the constant term is fairly important.

6. Evolution towards equilibrium

We have integrated numerically the master equation with an adaptative Runge-Kutta algorithm starting from configurations at thermal equilibrium at $40T$, where $T = 0.002$ is the phonon bath temperature, determining the evolution of the system. The system size is $N = 1000$. The occupation probability of the low energy configurations after a relaxation time $t = 10^5$ is represented in figure 6. The straight line is proportional to $\exp\{-E/kT\}$ and at equilibrium all configurations should lie in a line parallel to this one. We can appreciate that many configurations have arranged in lines with the slope of equilibrium. They correspond to configurations belonging to the same valley. We have reached equilibrium within valleys, but not between different valleys.

7. Conductivity

To study the conductivity of the system as it relaxes, we have integrated the master equation in the presence of an applied electric field. In this case we have considered eight diferent systems of size $N = 500$, at $T = 0.04$ and an electric field $F = T/20$. The conductivity as a function of time is plotted in figure 7 (red curve, left axis). We also plot the energy (blue curve, right axis) and note that their behaviours are quite different. The solid circle corresponds to the conductivity in equilibrium at the Maxwell time, i.e. the inverse of the conductivity in units of e^2/h. The conductivity decreases quickly at times shorter than the Maxwell time, when it reaches a value close to the equilibrium value and then relaxes slowly.

In figure 8 we represent the same conductivity data as in figure 7 on a logarithmic scale (red curve). The straight line has slope one and it represents a conductivity inversely proportional to the relaxation time. We have also plotted $-E^{-1/4}$ (blue curve, right axis), which should

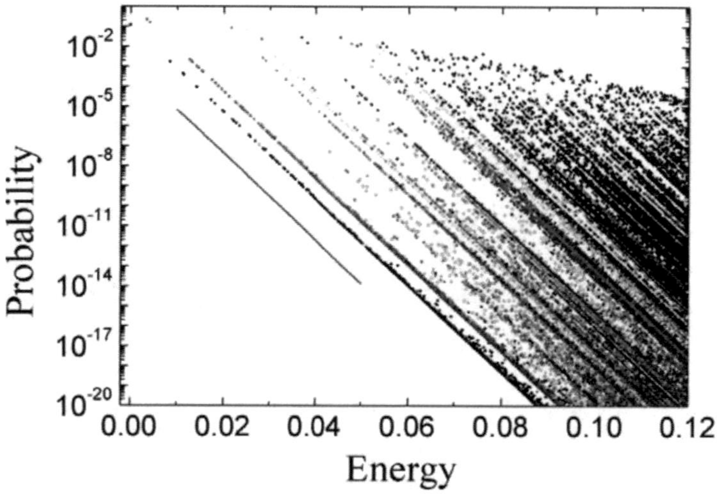

Figure 6. Occupation probability of low-energy configurations after the system has relaxed for a time $t = 10^5$. The straight line is proportional to $\exp\{-E/kT\}$.

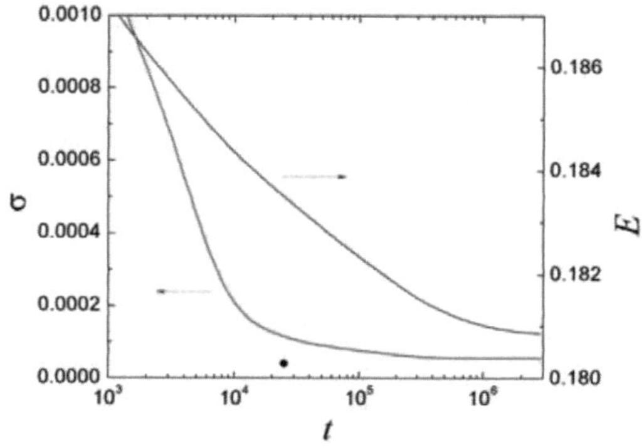

Figure 7. Conductivity as a function of time for eight different samples (dashed curves) and their average value (continuous curve) for $T = 0.04$. The solid circle corresponds to the conductivity in equilibrium at the Maxwell time.

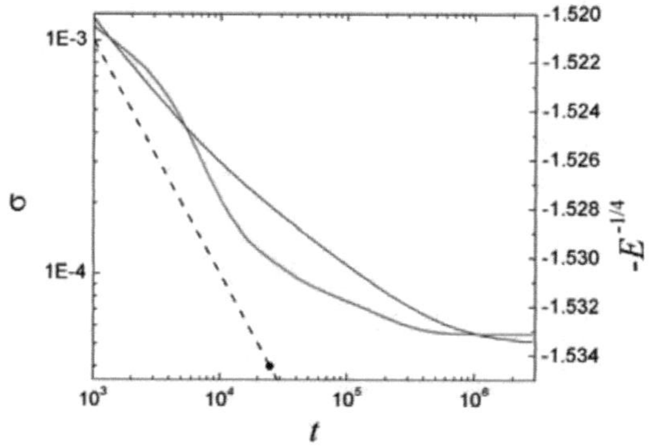

Figure 8. Conductivity as a function of time for eight different samples (dashed curves) and their average value (continuous curve) for $T = 0.04$. The solid circle corresponds to the conductivity in equilibrium at the Maxwell time

be roughly proportional to $-T_{\rm eff}^{-1/2}$, where $T_{\rm eff}$ is the effective temperature. If both curves in figure 8 would coincide, the conductivity would follow Efros and Shklovskii's law in terms of the effective temperature. We see that the curves are not very different, but the important featute of the change of behaviour at the Maxwell tiem is not capture by the energy relaxation.

8. Conclusions

The approach based on the master equation is very promising to study relaxation in electron glasses. To address the most interesting problems (shape of the dip, logarithmic relaxation, aging) we have to develop a renormalization technique adequate to the calculation of the conductivity. This is a problem much harder to solve than the previously developed renormalization approaches for the study of the energy.

Acknowledgments

References

[1] Z. Ovadyahu, Phys. Rev. Lett. **99**, 226603 (2007); –, Phys. Rev. Lett. **102**, 206601 (2009); Z. Ovadyahu and M. Pollak, Phys. Rev. B **68**, 184204 (2003).
[2] T. Grenet *et al.*, Eur. Phys. J. B **56**, 183 (2007); Delahaye, J. et al. Phys. Rev. Lett. **106**, 186602 (2011).
[3] L. M. Hernandez *et al.*, Phys. Rev. Lett. **91**, 126801 (2003).
[4] C. C. Yu, Phys. Rev. Lett. **82** 4074 (1999).
[5] E. Lebanon and M. Müller , Phys. Rev. B **72**, 174202 (2005).
[6] M. Pollak and Z. Ovadyahu, Phys. Stat. Sol. (c) **3**, 283 (2006).
[7] A. Glatz, V. M. Vinokur and Y. M. Galperin, Phys. Rev. Lett. **98**, 196401 (2007).
[8] D. N. Tsigankov *et al.*, Phys. Rev. B **68**, 184205 (2003); D. R. Grempel, Europhys. Lett. **66**, 854 (2004); A. B. Kolton, D. R. Grempel and D. Dominguez, Phys. Rev. B **71**, 024206 (2005).
[9] A. M. Somoza and M. Ortuño, Phys. Rev. B **72**, 224202 (2005).
[10] J. Kurchan, Nature **433**, 222 (2005).
[11] A.M. Somoza, M. Ortuño and M. Pollak, Phys. Rev. B **73**, 045123 (2006).
[12] A. B. Bortz, M. H. Kalos and J. L. Lebowitz, J. Comput. Phys. **17**, 10 (1975).
[13] A. Mobius and P. Thomas, Phys. Rev. B **55**, 7460 (1997).
[14] M. Ortuño, A. M. Somoza and M. Caravaca, Phys. Stat. Sol. (c) (in press).
[15] W. Krauth and O. Pluchery, J. Phys. A: Math. Gen. **27**, L715 (1994).
[16] A. Pérez-Garrido *et al.*, Phys. Rev. B **59** 5328 (1999).
[17] D. Menashe *et al.*, Phys. Rev. B **64**, 115209 (2001).
[18] J. G. Massey and M. Lee, Phys. Rev. Lett. **75** 4266 (1995).

Journal of Physics: Conference Series **376** (2012) 012008 doi:10.1088/1742-6596/376/1/012008

Screening and conductance relaxations in insulating granular aluminium thin films

Julien Delahaye and Thierry Grenet
Institut Néel and Université Joseph Fourier, B.P. 166, F-38042 Grenoble Cédex 9
E-mail: julien.delahaye@grenoble.cnrs.fr

Abstract. We have recently found in insulating granular Al thin film a new experimental feature [1], namely the existence of a conductance relaxation that is not sensitive to gate voltage changes. This conductance relaxation is related to the existence of a metallic-like screening in the film and can be used to estimate its characteristic length scale. In the present paper, we give some experimental details on how this feature was measured and present our first results on the screening length temperature dependence.

1. Introduction

Very slow conductance relaxations have been found during the last 20 years in some disordered insulators [2, 3, 4, 5, 6, 7]. They are characterized by two salient and related features: a slow decrease of the conductance after a quench of the samples at low temperature (4K) and a conductance "dip" in electrical field effect measurements. The current and still partial understanding of these features is based on the electron glass hypothesis [8, 9, 10]. According to theoretical studies [11, 12, 13, 14, 15, 16], a system of localized electrons with disorder and ill-screened interactions will need an infinite time to reach its equilibrium state at low temperature. The approach to equilibrium is associated with a decrease of the conductance, in qualitative agreement with the conductance relaxation observed after a quench at low temperature.

Let's precise what the conductance "dip" is with the example of Fig. 1. A MOSFET device which conducting channel is made of insulating granular Al thin film (see Fig. 2) was cooled down to $4.2K$ and maintained under a fixed gate voltage $V_{geq} = 0V$ for 2 hours. The gate voltage was then swept around V_{geq} at a constant scan rate and the $G(V_g)$ curve was recorded. A symmetrical dip, centered on V_{geq}, is clearly visible and reflects the partial equilibration of the system under V_{geq}. Out of the dip range, i.e for $|V_g| \geq 5V$, the conductance is roughly independent of the gate voltage and this baseline value G_{ref} can be used to defined the conductance dip amplitude $\Delta G = G_{ref} - G(V_{geq})$. G_{ref} is also called (wrongfully, see below) the "off equilibrium" conductance since the system has never been allowed to equilibrate at these V_g's. Contrary to our previous studies which were mainly focused on the time evolution of the conductance dip [6], we want to discuss here the time evolution of the baseline value G_{ref}. As we will see, it contains interesting information about the granular Al film properties.

2. Experimentals

The experimental challenge is to measure minute conductance variations of the granular Al channel at a fixed low T and over days. Since the relative amplitude of the conductance dip is

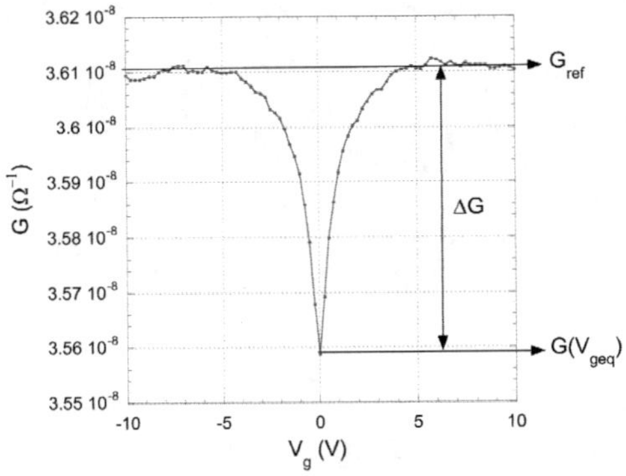

Figure 1. $G(V_g)$ curve measured on a granular Al film after a quench at $4.2K$. The gate voltage V_g was kept constant during 2 hours at $V_{geq} = 0V$ before sweeping V_g between $-15V$ and $+15V$. The film was $20nm$ thick and $R_\square = 550M\Omega$ at $4.2K$.

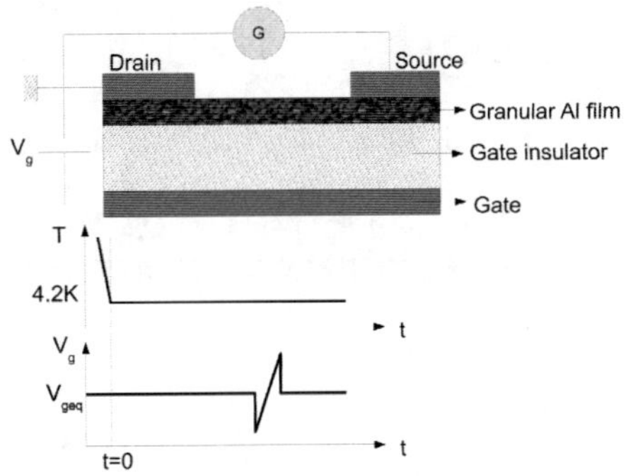

Figure 2. Schematic drawing of a MOSFET device made of granular Al thin film and the protocol used to obtained the $G(V_g)$ curve of Fig. 1.

of few percent or smaller, a relative precision of $\sim 10^{-2}\%$ is required.

The first problem is the T stability of the film during the measurement. Since the films are insulating, their resistance diverges exponentially when the T is lowered. mK temperature variations in the liquid He range correspond typically to few % changes in the conductance of the films, which is of the order of the variations we would like to measure as a function of time. The following set-up was thus employed. All the samples were mounted in a tight box and plunged into a $100l$ liquid He dewar. This box was either filled with some He exchange gas or let under vacuum. With He exchange gas, the quench to $4.2K$ was fast ($\simeq 10min$ from $90K$) but then the film follows the mK variations of the liquid He bath around $4.2K$. A carbon glass thermometer close by was used to correct these T variations. Under vacuum, the quench to low T was much longer (few hours), but it was then possible to regulate the temperature with a high precision (better than $0.2mK$) by using the carbon glass thermometer and a resistive heater. The available temperatures in that case are in the range $5 - 15K$.

The second problem is the sensitivity of the electronic devices (lock-in, current amplifier, etc.) to room temperature drifts. For example, the current amplifier gain in the range 10^6 - 10^8 has a T drift of $100ppm/^oC$. We specially built a thermalized chamber with a T stability better than $0.1K$ in which all the electronic devices were placed. Doing so, we were able to detect conductance variations as small as $10^{-3}\%$ over weeks of measurements.

We have exclusively used granular Al films made by electron gun evaporation of Al under a partial pressure of O_2, as described elsewhere [6]. In order to build MOSFET devices, granular Al films were deposited on top of heavily doped Si wafers (the gate) covered by a $100nm$ thick thermally grown SiO_2 layer. The film conductance was measured by using a two terminal AC technique, employing a FEMTO current amplifier DLPCA 200, and a lock-in amplifier SR 7265. The source-drain voltage was such that G stays in the ohmic regime.

3. The baseline relaxation

In order to measure the baseline G_{ref} time variations, we follow a simple experimental protocol. The MOSFET devices are cooled down from room T to liquid He (or to a T slightly above) and once the T has reached the desired value, we measure $G(V_g)$ curves around V_{geq} at constant time interval. The time between two sweeps is kept much larger than the sweep time, usually $6000s$ between two $G(V_g)$ curves and $250s$ for a $-15V/+15V$ sweep.

A typical result for a $20nm$ thick film quenched with He exchange gas is shown in the Fig. 3 below. Without any T corrections, the observed baseline conductance variations of few % mainly reflect the mK fluctuations of the He bath. Interestingly, the dip amplitude is less sensitive to T so that T variations are small enough to allow a measurement of its time development. This is illustrated in the Fig. 4 where ΔG is plotted as a function of the time elapsed after the quench. The $\ln(t)$ dependence reported and discussed in previous studies [6] is clearly visible.

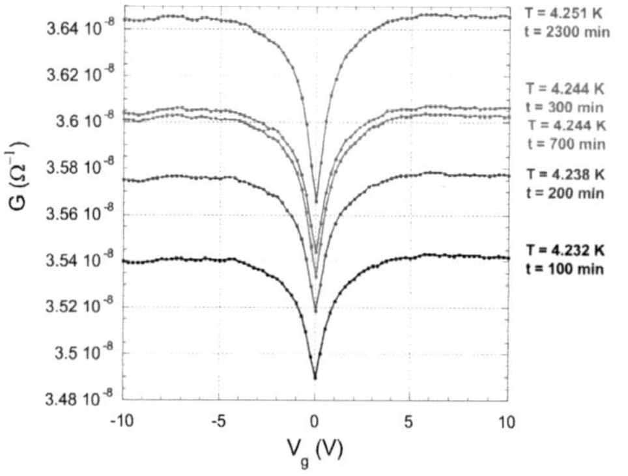

Figure 3. $G(V_g)$ curves measured at different times t after a cool down to $4.2K$, without any corrections of the He bath T fluctuations. $V_{geq} = 0V$ between V_g sweeps. Same film as in Fig. 1.

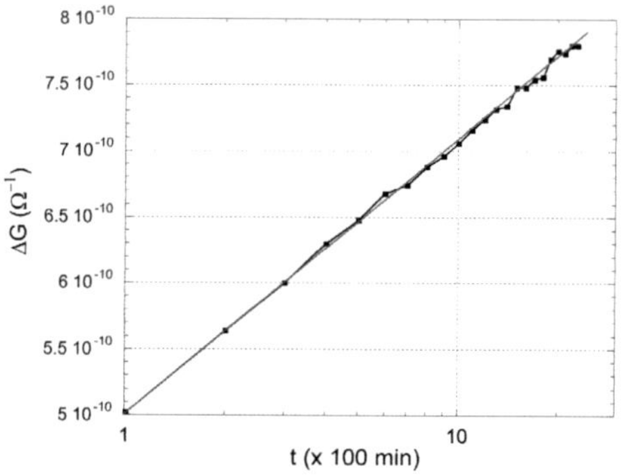

Figure 4. Corresponding amplitude of the conductance dip $\Delta G(t) = G_{ref} - G(V_{geq})$.

If we now correct for the T variations of the He bath (we can measure after a long time the G-T dependence), we get the curves shown in Fig. 5 and 6. Both the baseline and the conductance minimum at V_{geq} decrease with a $\ln(t)$ dependence. It will be noted that the curve of the Fig. 4 is simply the difference of the two $\ln(t)$ curves of the Fig. 6.

We checked that the V_{geq} values has no effect on the baseline decrease. Quantitatively similar G_{ref} relaxations are observed in Fig. 7 and 8 when V_{geq} was first set to $V_{geq1} = -7.5V$ and then change to $V_{geq2} = +7.5V$ 800 min after a quench to $5.3K$. Two dips are visible in the $G(V_g)$ curves, but for V_g values far enough from V_{geq1} and V_{geq2}, a regular $\ln(t)$ decrease of the conductance is obtained. During this run of measurement, we have also suspended the V_g sweeps during 30 hours. No significant change in the baseline decrease is found which exonerates the V_g sweeps in the G relaxations.

4. Discussion

The $\ln(t)$ dependence of $G(V_{geq})$ reflects the well known approach of the system to equilibrium after a quench [3, 6]. But how can we understand the $\ln(t)$ relaxation of the baseline? In Ref. [1], we described in details how a metallic-like screening length L_{sc} smaller than the thickness T_h of the films ($20nm$ for the previous figures) can provide a simple explanation to the observed baseline relaxation. For a given V_g, only the part of the film which is at a distance smaller than L_{sc} from the gate insulator will feel an electric field (let's call it the "unscreened" layer). The field in the other part of the film (the "screened" layer) will be much smaller. If the screening is much faster than the typical time scale of the V_g changes ($\simeq 1s$), there won't be any electrostatic

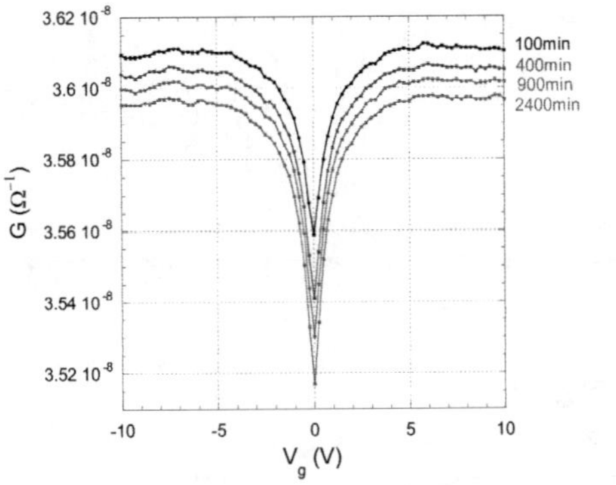

Figure 5. $G(V_g)$ curves measured at different times after a cool down to $4.2K$ with a correction of the T variations. Same film as in Fig. 1.

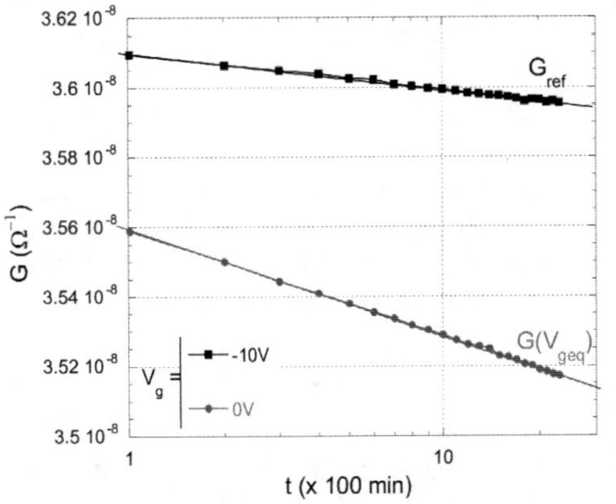

Figure 6. Corresponding $G(t)$ curves at V_{geq} ($V_{geq} = 0V$) and out of the conductance dip region ($V_g = -10V$).

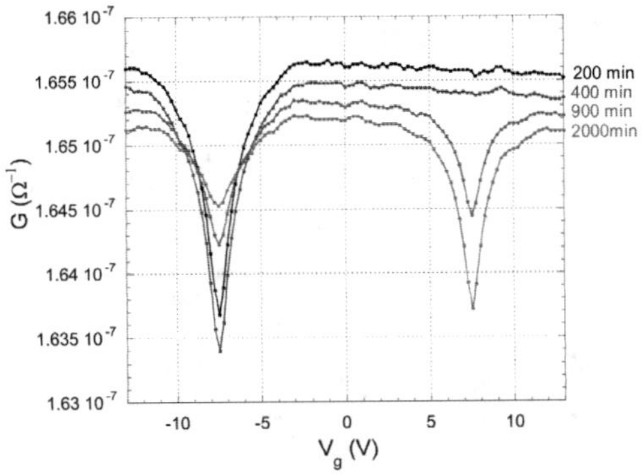

Figure 7. $G(V_g)$ curves measured at different times after the cool down at $5.3K$. V_{geq} was first set to $V_{geq1} = -7.5V$ and changed to $V_{geq2} = +7.5V$ after 800 min. Same film as in Fig. 1.

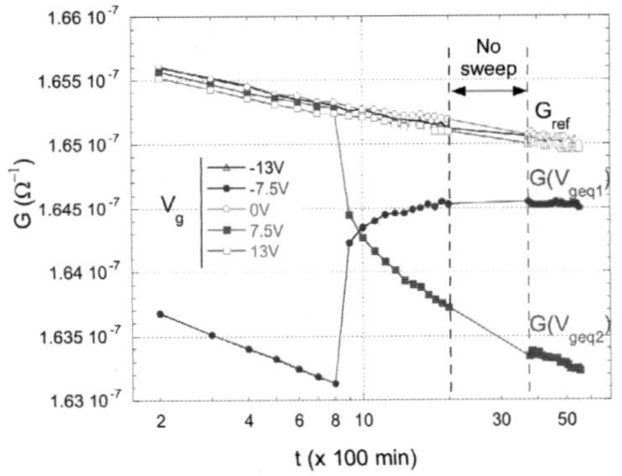

Figure 8. Corresponding $G(t)$ curves at different V_g values. The V_g sweeps were suspended during 29 hours, 33 hours after the cool down.

perturbation during a V_g sweep in the "screened" layer. In other words, only the "unscreened" layer is pushed out of "equilibrium" by the V_g sweeps. By assuming a parallel conduction for the different layers, we can say that the $\ln(t)$ dependence of $G(V_{geq})$ reflects the relaxation of the whole film ("screened" and "unscreened" layers) towards equilibrium, whereas the $\ln(t)$ dependence of G_{ref} reflects the relaxation of the "screened" layer alone (the relaxation of the "unscreened" layer is given by the conductance dip growth). It was shown in Ref. [1] that the slope ratio SR of the $G(V_{geq}) - \ln(t)$ slope to the $G_{ref} - \ln(t)$ slope is given by:

$$SR = T_h/(T_h - L_{sc}) \tag{1}$$

and can thus be used to estimate the L_{sc} value. In the Fig. 6, SR is equal to 2.3, which gives $L_{sc} = 13nm$ ($T = 4.2K$). We have measured this slope ratio at $4.2K$ on other 20nm thick

Journal of Physics: Conference Series **376** (2012) 012008 doi:10.1088/1742-6596/376/1/012008

granular Al thin films with square resistances $R_\square$ between $1M\Omega$ and $10G\Omega$. The slope ratios were all found between 2 and 3, corresponding to quite similar L_{sc} values in the range $10-13nm$. This L_{sc} estimate of $\simeq 10nm$ at $4K$ was confirmed by the study of the $G(V_g)$ relaxations for granular Al films of different thicknesses [1]. It lies between the percolation radius L_0, estimated to be about $20-40nm$ at $4.2K$ (see below), and the typical grain size of Al, which according to TEM studies is in the range $2-4nm$.

The existence of a metallic-like screening length in our insulating films is not surprising. The films have a low but finite conductivity at low temperature. Some electrons can thus move by hopping through the whole sample and screen an electric field. Other electrons that remain localized during the time scale of the experiment give a dielectric-like screening contribution (polarization). A transition between a metallic screening at high T, where most of the electrons are diffusive, and a slow dielectric response at low T, where most of the electrons remain localized in finite size clusters was indeed found in numerical studies of disordered insulators [18, 19, 13].

One may expect that the frequency and the temperature dependence of the screening properties are different in our hopping system from the metal case. By measuring the slope ratio SR at different temperatures between $4.2K$ and $10K$ for the 20nm thick film of the previous figures (see Fig. 9 and 10), we found that L_{sc} decreases as the T (and thus G) increases. This increase can be described by a $1/T$ dependence predicted by some theoretical studies [20], but the uncertainty is large.

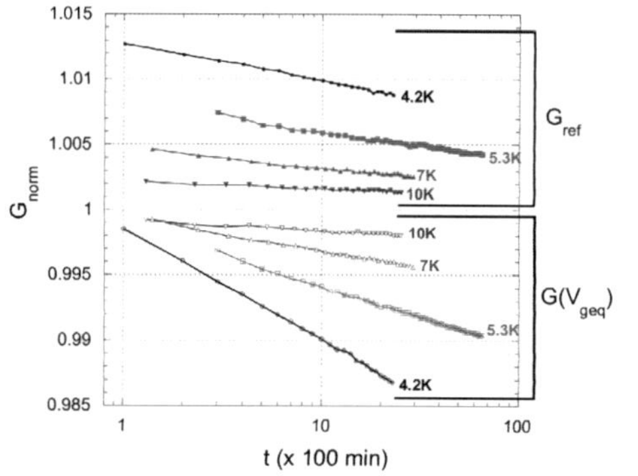

Figure 9. Normalized conductance relaxations measured at $-10V$ (G_{ref}) and at $V_{geq} = 0V$ ($G(V_{geq})$) as a function of time and after the cool down at different temperatures. Same film as in Fig. 1.

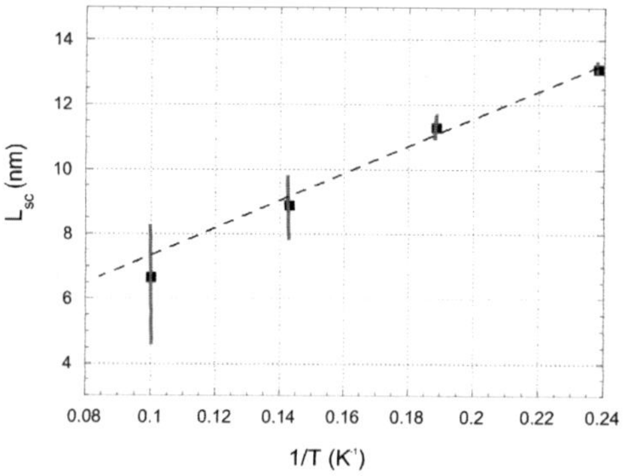

Figure 10. The screening length values extracted from the slope ratios SR of Fig. 9 plotted versus $1/T$.

We do not see any evidence for a large change with time in the effective screening length over the time scales probed by our experiment (from $1s$ to few days). Since glassy effects represent only conductance changes of few percent or less, their contribution to the screening might also be small and not detectable by our experiments. According to theoretical studies of the electron glass, the single electrons hops are believed to be fast (Maxwell time) and could be responsible for the short time screening length measured here. Much slower multi-electron hops will result in a slow decrease of the screening length as a function of time [20].

In light of these findings, some of the results that were published before have to be quantitatively reconsidered. It is indeed the case of our percolation radius (L_0) estimates obtained from the conductance fluctuations of $20nm$ thick mesoscopic films [17]. Since only the film layer located at a distance smaller than L_{sc} to the gate insulator is sensitive to V_g

changes, the L_0 values published in Ref. [17] must thus be divided by $\sqrt{(T_h/L_{sc})} = \sqrt{2}$. Due to the T dependence of the screening length highlighted in Fig. 10, the actual decrease of the relative conductance dip amplitude $\Delta G/G$ as T is increased is weaker than it was published in Ref. [6].

Finally, we should say that no baseline relaxation was found in indium oxide films where similar measurements to ours have been made. Published data concern films of 20nm thick or less, and this will suggest a screening length larger than 20nm. It would thus be interesting to measure thicker films to see if the screening length limit can be attained. Such experiments are obviously not possible for ultrathin films of metals, in which the insulating character is due to their thinness.

5. Conclusion

In conclusion, we have demonstrated in our granular Al insulating films the existence of a conductance relaxation which is undisturbed by a gate voltage change. This relaxation is observed after a quench at low temperature and obeys a $\ln(t)$ dependence. According to Ref. [1], the comparison with the $\ln(t)$ increase of the conductance dip amplitude allows an estimate of the metallic-like screening length of the films. This length is found to be of the order of $10nm$ at $4K$ and to decrease as the temperature is increased.

Acknowledgments

This research has been supported by the French-National Research Agency ANR (Contract 05-JC05-44044), the Région Rhônes Alpes (CIBLE 2010 program) and the Université Joseph Fourier (SMINGUE 2010 program). Discussions with A. Frydman and M. Pollak are gratefully acknowledged.

References

[1] Delahaye J, Honoré J and Grenet T 2011 *Phys. Rev. Lett.* **106** 186602
[2] Ben-Chorin M, Kowal D and Ovadyahu Z 1991 *Phys. Rev. B* **44** 3420
[3] Vaknin A, Ovadyahu Z and Pollak M 2002 *Phys. Rev. B* **65** 134208
[4] Martinez-Arizala G, Grupp D E, Christiansen C, Mack A M, Markovic N, Seguchi Y, Goldman A M 1997 *Phys. Rev. Lett.* **78** 1130
[5] Grenet T 2003 *Eur. Phys. J. B* **32** 275
[6] Grenet T, Delahaye J, Sabra M and Gay F 2007 *Eur. Phys. J. B* **56** 183
[7] Frydman A, this conference
[8] Davies J H, Lee P A and Rice T M 1982 *Phys. Rev. Lett.* **49** 758
[9] Grünewald M, Pohlmann B, Schweitzer L and Wurtz D 1982 *J. Phys. C* **15** L1153
[10] Pollak M and Ortuño M 1982 *Sol. Energy Mater.* **8** 81
[11] Tsigankov D N, Pazy E, Laikhtman B D and Efros A L 2003 *Phys. Rev. B* **68** 184205
[12] Malik V and Kumar D 2004 *Phys. Rev. B* **69** 153103
[13] Kolton A B, Grempel D R and Dominguez D 2005 *Phys. Rev. B* **71** 024206
[14] Lebanon E and Müller M 2005 *Phys. Rev. B* **72** 174202
[15] Somoza A M, Ortuño M and Caravaca M 2008 *Phys. Rev. Lett.* **101** 056601
[16] Amir A, Oreg Y and Imry Y 2009 *Phys. Rev. Lett.* **103** 126403
[17] Delahaye J, Grenet T and F. Gay 2008 *Eur. Phys. J. B* **65** 5
[18] Xue W and Lee P A 1998 *Phys. Rev. B* **38** 9093
[19] Díaz-Sanchez A, Ortuño M, Pollak M, Peréz-Garrido and Mobius M 1999 *Phys. Rev. B* **59** 910
[20] Müller M and Pankov S 2007 *Phys. Rev. B* **75** 144201

14th International Conference on Transport in Interacting Disordered Systems (TIDS-14) IOP Publishing
Journal of Physics: Conference Series **376** (2012) 012009 doi:10.1088/1742-6596/376/1/012009

Elementary excitations and avalanches in the Coulomb glass

Matteo Palassini and Martin Goethe

Departament de Física Fonamental, Universitat de Barcelona, Av. Diagonal 645, E–08028 Barcelona, Spain.

E-mail: palassini@ub.edu, MartinGoethe@ub.edu

Abstract. We study numerically the statistics of elementary excitations and charge avalanches in the classical Coulomb glass model of localized charges with unscreened Coulomb interaction and disorder. We compute the single-particle density of states with an energy minimization algorithm for systems of up to 100^3 sites. The shape of the Coulomb gap is consistent with a power-law with exponent $\delta \simeq 2.4$ and marginally consistent with exponential behavior. The results are also compared with a recently proposed self-consistent approach. We then analyze the size distribution of the charge avalanches produced by a small perturbation of the system. We show that the distribution decays as a power law in the limit of large system size, and explain this behavior in terms of the elementary excitations. Similarities and differences with the scale-free avalanches observed in mean-field spin glasses are discussed.

1. Introduction

In this contribution, we study two related aspects of the physics of disordered systems of localized charges with long-range interaction. In Sec. 2, we reconsider the long-standing puzzle of the exponential suppression of the one-particle density of states in three dimensions, predicted long ago [1, 2] but never observed in numerical studies. In Sec. 3, we investigate the size distribution of the charge rearrangements, or avalanches, induced by a small displacement of a charge.

We consider the Coulomb glass model [1], in which KN charges occupy $N = L^d$ sites arranged in a regular d-dimensional lattice of linear size L. The Hamiltonian in dimensionless units is

$$\mathcal{H} = \sum_{i<j}^{N} (n_i - K)\frac{1}{r_{ij}}(n_j - K) + \sum_{i=1}^{N} n_i \varphi_i \,, \tag{1}$$

where $n_i \in \{0, 1\}$ is the occupation number for site i, and $\sum_{i=1}^{N} n_i = KN$. The φ_i's are independent, identically distributed Gaussian random variables with zero mean and standard deviation W.

2. Elementary excitations

As first pointed out by Pollak [3] and Srinivasan [4], the one-particle density of states (1-DOS), $g(\varepsilon) = N^{-1}\sum_{i=1}^{N}\delta(\varepsilon - \varepsilon_i)$, where $\varepsilon_i = \sum_{j\neq i}^{N}(n_j - 1/2)/r_{ij} + \varphi_i$, is depleted near the Fermi level ε_F due to the long-range interaction. Efros and Shklovskii [5] famously showed that the stability

Published under licence by IOP Publishing Ltd

of the ground state against one-particle hops requires $g(\varepsilon)$ to vanish at ε_F (creating a so-called Coulomb gap) and to satisfy the bound

$$g(\varepsilon) \leq c_d|\varepsilon - \varepsilon_F|^{d-1} \tag{2}$$

near ε_F, where the constant c_d depends on d. Efros [1] and Baranovskii et al. [2] considered a self-consistent equation (SCE), that for $K = 1/2$ (which sets $\varepsilon_F = 0$) and $d = 3$ reads

$$g(\varepsilon) = g_0(\varepsilon)\exp\left[-\frac{2\pi}{3}\int_0^\infty \frac{g(\varepsilon')d\varepsilon'}{(\varepsilon' + |\varepsilon|)^3}\right], \tag{3}$$

where $g_0(\varepsilon)$ is the bare 1-DOS. For asymptotically small ε, the solution to Eq.(3) is $g(\varepsilon) = (3/\pi)\varepsilon^2$, i.e. the bound in Eq.(2) is saturated. For $d = 2$, an analogous SCE gives asymptotically $g(\varepsilon) = (2/\pi)|\varepsilon|$. The behavior $g(\varepsilon) = c_d|\varepsilon - \varepsilon_F|^{d-1}$ is universal in the sense that it does not depend on the bare 1-DOS.

Due to the Coulomb gap, the low energy elementary excitations can be roughly divided [2] in *charged excitations* (adding or removing a charge from a site with $|\varepsilon_i| < \Delta$, where Δ is the gap width) and *dipole excitations*, i.e. electron-hole pairs at nearby sites i, j with $(\varepsilon_j, -\varepsilon_i) > \Delta$ and excitation energy $\omega = \varepsilon_j - \varepsilon_i - 1/r_{ij} \ll (\varepsilon_j, -\varepsilon_i)$. If the dc conductivity is dominated by long one-particle hops, and the bound in Eq.(2) is saturated, the variable-range hopping $T^{1/2}$ law follows [5, 1]. The dipole density of states (2-DOS) $\phi(\omega)$ vanishes only logarithmically as $\omega \to 0$ in $d = 3$ [2, 6], thus low energy dipoles dominate ac conductivity and the specific heat.

Equation (2) does not consider the stability of the ground state against multi-particle hops, so an important question is whether these produce a harder gap in $g(\varepsilon)$. In Refs.[1, 2] it was argued that, for $d = 3$, the stability against a charge excitation of energy ε and simultaneous excitation of dipoles at distance $r < \varepsilon^{-1/2}$ from it requires $g(\varepsilon)$ to vanish exponentially as $g(\varepsilon) \propto P(\varepsilon)$ for $\varepsilon \ll \Delta$, where $P(\varepsilon)$ is given by Eq.(4) with $f(x) = 1$ and $\gamma = 1.5$. For $d = 2$, this criterion does not produce a hardening of the linear gap. Contrary to these results, recent mean-field studies [7] predict a quadratic $g(\varepsilon)$ for $d = 3$, by connecting the shape of the gap to a putative glass transition. Subsequent numerical studies [8, 9, 10], however, provided evidence against the existence of such a transition.

Numerical computations of the 1-DOS in $d = 3$ have failed so far to observe an exponential gap, and generally favor a power law $g(\varepsilon) = c_d|\varepsilon|^\delta$ with $\delta \geq 2$ [8, 11, 12, 13, 14]. Recently, Efros, Skinner, and Shklovskii (henceforth referred to as ESS) [15] proposed a modified SCE in which $g_0(\varepsilon)$ in Eq.(3) is replaced by $g_0(\varepsilon)P(\varepsilon)$, where

$$P(\varepsilon) = \exp[-\gamma\, f(\Delta/\varepsilon)\,(\Delta/\varepsilon)\,\ln^{-7/4}(\Delta/\varepsilon)] \tag{4}$$

and the crossover function $f(x) = (1-x)^\eta\theta(1-x)$, with $\eta > 7/4$, is introduced to interpolate between the $\varepsilon \ll \Delta$ regime and $\varepsilon \simeq \Delta$ where $P(\varepsilon) \simeq 1$. The solution to this modified SCE gives asymptotically $g(\varepsilon) \sim g_0(\varepsilon)P(\varepsilon)$ for $\varepsilon \ll \Delta$ (a non-universal behavior since Δ enters explicitly), and a softer behavior for $\varepsilon \to \Delta$. ESS argue that this "delayed" onset of the exponential hardening is the reason why it has not been observed in numerical simulations.

2.1. Numerical results

We consider the Hamiltonian in Eq.(1) for $d = 3$ and $K = 1/2$ with periodic geometry, namely we surround the L^3 simulation cell with an infinite number of identical images, and sum over the interactions between site i of the central cell and site j of all the images using the Ewald method [10]. This gives an effective interaction $r_{ij}^{-1} + v(\mathbf{r}_{ij})$, where $v(\mathbf{r}_{ij})$ is a correction of order L^{-1} [16]. For each disorder realization (sample) $\{\varphi_i\}$, we start from a random configuration and perform energy-decreasing one-particle hops chosen uniformly at random among all unstable electron-hole pairs, until we reach a configuration stable against all one-particle hops, or *pseudo ground state* [17]. To this end, we use a modified version of the algorithm of Ref. [18] adapted to the periodic geometry [16].

We report here our results for $L = 100$. We simulate 32 (96) samples for $W = 2$ ($W = 4$). To reduce the sample-to-sample fluctuations of ε_F, which tend to fill the Coulomb gap, before averaging over samples we shift $g(\varepsilon)$ by $(\varepsilon_a + \varepsilon_b)/2$ for each sample, where ε_a and ε_b are such that $\int_{-\infty}^{\varepsilon_a} d\varepsilon\, g(\varepsilon) = \int_{\varepsilon_b}^{\infty} d\varepsilon\, g(\varepsilon) = p$. In Fig.1 we plot $g(\varepsilon)W$ as a function of $\varepsilon W^{1/2}$, excluding energies affected by finite-size effects ($\varepsilon \leq 0.015$, as we determined from a finite-size scaling analysis [16]). For $0.1 < \varepsilon W^{1/2} < 0.4$ (or $0.07 < \varepsilon/\Delta < 0.26$, where we define $\Delta = (g_0(0)\pi/3)^{1/2} = (18/\pi)^{1/4}/W^{1/2}$) the data are close to $g(\varepsilon) = (3/\pi)\varepsilon^2$, but a deviation is visible for small ε. We solved the modified SCE with the same numerical method and parameters ($\eta = 4, \gamma = 1.5$) as ESS for $g_0(\varepsilon) = (2\pi W^2)^{-1/2} \exp[-\varepsilon^2/(2W^2)]$. (For the box distribution, our solution agrees with theirs. Incidentally, we also solved numerically Eq.(3), finding oscillations in ε around the asymptotic solution $3\varepsilon^2/\pi$ [19]). As shown in Fig.1, the solution for $W = 2$ is fairly close to the data at intermediate energies, but for small ε it gives a steeper behavior. Taking $\eta = 3$ improves only slightly the agreement. For small ε the data for $W = 2, 4$ are superimposed, consistent with the asymptotic scaling $g(\varepsilon) \sim g_0(\varepsilon)P(\varepsilon)$, but are actually far from the function $g_0(\varepsilon)P(\varepsilon)$, showing that much smaller energies are needed to see the exponential hardening.

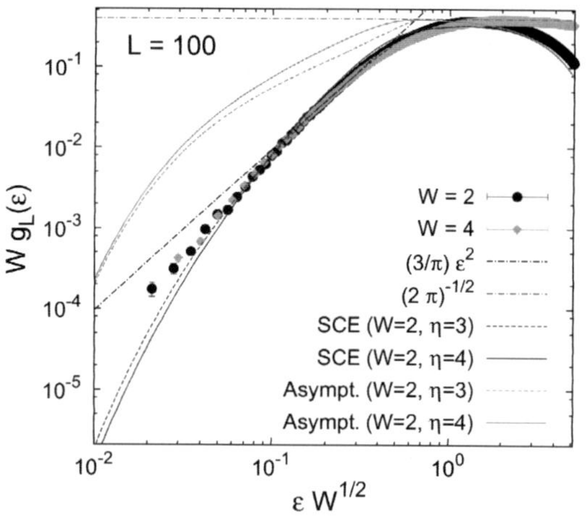

Figure 1. 1-DOS for two different disorder strengths W, averaged with shifting parameter $p = 0.499$. The curved lines represent the numerical solution of the modified SCE and the asymptotic behavior $g_0(\varepsilon)P(\varepsilon)$.

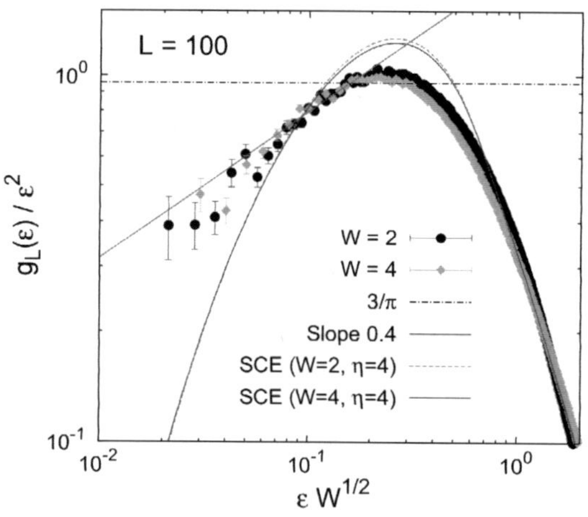

Figure 2. Same as Fig.1 with a different scaling. Note that the solutions of the modified SCE for $W = 2$ and 4 are almost indistinguishable.

The deviation from both the quadratic behavior and the modified SCE becomes very clear by plotting $g(\varepsilon)/\varepsilon^2$, see Fig.2. The data are also consistent with $g(\varepsilon) = c_d|\varepsilon|^\delta$ with $\delta \simeq 2.4$ for $\varepsilon W^{1/2} < 0.2$, displayed with the sloped line in Fig.2. In fact, the data for different W are well described by $g(\varepsilon) = a_d|\varepsilon|^\delta W^{1-\delta/2}$ with $\delta \simeq 2.4, a_d \simeq 2$.

To estimate the systematic error due to pseudo ground states not being true ground states, we ran the algorithm 20 times per sample for a subset of 196 samples with $L = 60$, $W = 2$. The sample-averaged $g(\varepsilon)$ computed with the pseudo ground states a and b with lowest and highest energy for each sample agree within the error bars. In average, a and b differ in only $\simeq 1\%$ of the sites, and their relative energy difference is very small ($\simeq 2.5 \cdot 10^{-5}$, a factor 10^2 smaller than the sample-to-sample energy fluctuation). Hence, it seems unlikely that this systematic error affects our conclusion. We report elsewhere [16] these tests and a detailed finite-size scaling analysis.

In an earlier work [8] we found evidence for a quadratic gap from a scaling analysis of the temperature and energy dependence of the 1-DOS obtained with equilibrium Monte Carlo. While those results are unaffected by the above systematic error (equilibration was carefully checked), they are limited to $L \leq 10$, so because of the finite-size effects below $\varepsilon \sim L^{-1}$ we could only explore the energy range in which the data in Figs.1,2 are approximately quadratic. In this energy range the present data and those of Ref.[8] agree quantitatively, despite being obtained with completely different algorithms.

In conclusion, the data from pseudo ground states show a clear deviation from a quadratic gap and are consistent, in the energy range reached, with a power-law gap with exponent ≥ 2.4, in agreement with Refs.[11, 12]. The data also agree qualitatively but not quantitatively with the scenario of ESS. Much larger systems will be needed to discriminate between an exponential and a power-law gap.

3. Charge avalanches

Let us now consider the following numerical experiment. Starting from a pseudo ground state found as in Sec.2, we perturb it by inserting an extra charge or by exciting a dipole. This will in general destabilize some electron-hole pairs. We then relax one of the unstable pairs, which in turn creates new unstable pairs, and continue in this way until we stop upon reaching a new pseudo ground state, after a number S of hops. Avalanche processes of this kind have been studied in disparate systems such as earthquakes, sandpiles, and Barkhausen noise in magnets [20]. A well studied theoretical example is the random field Ising model (RFIM). The RFIM Hamiltonian is identical to Eq.(1) when we take $K = 1/2$ and truncate the interaction to nearest neighbors. In Ref.[21] avalanches were triggered by a small uniform external field that destabilizes only one spin, and evolved via zero-temperature single spin-flip dynamics. At a critical value of the disorder $W = W_c$, the probability distribution of the avalanche size was found to decay as

$$p(S) \sim S^{-\tau} \exp(-S/S_c) \tag{5}$$

for large S, where $\tau = 1.60 \pm 0.06$ and S_c is a cutoff that diverges with the system size. Hence, $p(S)$ is scale-free in the thermodynamic limit, $L \to \infty$. For $W > W_c$, the cutoff tends to a finite value $S_c \to S^*$ for large L, with $S^* \to \infty$ as $W \to W_c$. For $W < W_c$ there are system-size avalanches with finite probability [21].

Pazmandi et al. [22] simulated the same dynamics in the infinite-range Sherrington-Kirkpatrick (SK) spin glass model and found a power-law behavior with a cutoff proportional to the system size, and estimated $\tau \simeq 1$. Unlike in the RFIM, however, this did not require fine-tuning of parameters (the criticality is "self-organized" [22]). Recently, the authors of Ref.[23] related the equilibrium avalanches in the SK model to the marginal criticality of its equilibrium glass phase, and obtained the analytical result $p(S) \sim S^{-\tau}$ with $\tau = 1$. It was suggested that a scenario similar to that of the SK model might be at play in the Coulomb glass [24, 23]. Although a glass phase was ruled out numerical studies down to very low temperatures [8, 9, 10], the possibility remains that the system is critical at zero temperature. It is thus interesting to ask whether the avalanches are scale-free. Avalanches in the Coulomb glass provide a mechanism for nonlinear screening [25, 26], which is relevant for capacitance and conductance experiments on disordered insulators and granular metals [27, 28].

Figure 3 shows $p(S)$ from the numerical experiment described above for $d = 3$, $W = 2$, and $K = 1/2$. The avalanches are triggered by exciting the lowest energy dipole that produces an instability and are evolved by random hops as in Sec.2. The most important feature of Fig.3 is the exponential cutoff that increases linearly with L. As shown in Fig.4, the data can be rescaled according to Eq.(5) with $\tau \simeq 1.5$ and $S_c = aL$, $a \simeq 0.48$. A form $p(S) \sim S^{-\tau} \exp[-(\frac{S}{aL})^2]$ also

Journal of Physics: Conference Series **376** (2012) 012009 doi:10.1088/1742-6596/376/1/012009

fits reasonably well $\tau \simeq 1.7$ and $S_c \propto L$. This scale-free behavior (extrapolating to $L \to \infty$) is robust and does not require fine tuning. In fact, we observe a divergent cutoff for other values of K near 1/2, when hops are performed in order of increasing hopping length instead of uniformly at random, and even for huge disorder ($W = 32$).

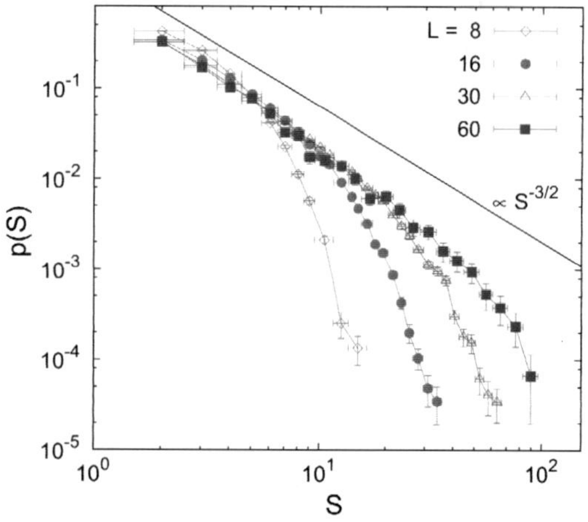

Figure 3. Avalanche size distribution for dipole-triggered avalanches. The data are averaged over 20000, 48000, 29000, 2135 samples for $L = 8, 16, 30, 60$ respectively.

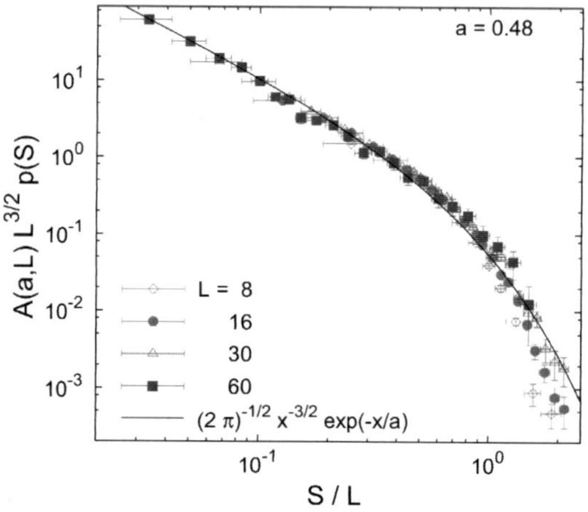

Figure 4. Scaling plot of the avalanche size distribution assuming Eq.(5) with $\tau = 3/2$ and $S_c = aL$. $A(a, L)$ is a normalization factor slowly varying with L, computed analytically from the branching process model.

Although these results may suggest an analogy with the self-organized critical avalanches of the SK model, we argue that they can be understood more simply in terms of the elementary charged and dipole excitations, without invoking a critical glass phase. We repeated our numerical experiment forbidding hops longer than a certain fixed length R_{max} independent of L, and found that in this case $p(S)$ has an exponential cutoff that does not increase with L [16]. If we choose R_{max} proportional to L, we find again $S_c \propto L$, which demonstrates that scale-free avalanches require arbitrarily large hops. In a glass phase the critical avalanches arise due to an infinite correlation length [7, 23], thus in our understanding the scale-free behavior should be present also for finite R_{max}, contrary to our results.

We can understand qualitatively how scale-free avalanches arise by considering the response to dipole and charge excitations, neglecting correlations among successive hops. When a dipole is excited, a number $O(\phi(0) \log L)$ of dipoles become unstable due to the $1/r^3$ dipole-dipole interaction (we neglect here the logarithmic gap of the 2-DOS $\phi(\omega)$). These are mostly near the excited dipole (the probability to find an unstable dipole at distance r decays as $1/r$), thus the avalanche cannot propagate far if only short hops are allowed. In addition to the dipoles, a certain number of arbitrarily long pairs also become unstable. A long hop can create a charged excitation in a region still untouched by the avalanche. The charged excitation destabilizes a number $O(\phi(0)L)$ of dipoles. When one of these dipoles flips it creates, as we just argued, a local finite avalanche and, with finite probability, new long unstable pairs. This process can be thought of as a branching process with branching ratio unity, which gives critical avalanches with $\tau = 3/2$. The above argument also explains why S_c is linear in L. In fact, the cutoff is determined by long hops that reach the boundary of the system. Since a long hop creates a number $O(L)$ of subsequent hops, $S_c \propto L$ follows. By the same argument, for $d = 2$ the cutoff scales as $S_c \propto \log L$, which indeed we observe numerically [16].

An avalanche exponent $\tau = 3/2$ is typical of mean-field models [29]. In this spirit, the branching process model considers the avalanche as a collection of independent events. Let us just mention that in this model, $p(S)$ for avalanches triggered by a charge injection (instead of a dipole excitation) is given by a convolution of M Galton-Watson processes, corresponding to subavalanches initiated by dipoles made unstable by the charge injection, where M is a Poisson variable with $O(L)$ mean, and fits well our numerical results [19].

We conclude by noting that in a physical dynamics with distance-dependent hopping probability, time introduces a cutoff on the maximal hopping length, and thus on the power-law tail of $p(S)$. We expect that a nonzero temperature produces a similar effect. Avalanches at finite temperature could be studied with a "damage spreading" algorithm [30].

Acknowledgments

We thank Alexei Efros, Arnulf Möbius, Markus Müller, Boris Shklovskii, and Brian Skinner for useful discussions. The work is supported by the Ministerio de Ciencia e Innovación (FIS2009-13730-C02-01, AP2007-01005). The computations were performed on the Spanish Supercomputing Network (RES) and the Albeniz cluster at UB.

References

[1] A. L. Efros, J. Phys. C **9**, 2021 (1976).
[2] S. D. Baranovskii, B. I. Shklovskii, A. L. Efros, Sov. Phys. JETP **51**, 199 (1980).
[3] M. Pollak, Disc. Faraday Soc. **50**, 13 (1970).
[4] G. Srinivasan, Phys. Rev. B **4**, 2581 (1971).
[5] A. L. Efros and B. I. Shklovskii, J. Phys. C **8**, L49 (1975).
[6] E. Cuevas, R. Chicón, and M. Ortuño, Physica B **160**, 293 (1990).
[7] M. Müller and L. B. Ioffe, Phys. Rev. Lett. **93** (2004); S. Pankov and V. Dobrosavljević, Phys. Rev. Lett. **94**, 046402 (2005); M. Müller and S. Pankov, Phys. Rev. B **75**, 144201 (2007).
[8] M. Goethe and M. Palassini, Phys. Rev. Lett. **103**, 045702 (2009).
[9] B. Surer, H. G. Katzgraber, G. T. Zimanyi, B. A. Allgood, and G. Blatter, Phys. Rev. Lett. **102**, 067205 (2009).
[10] M. Goethe and M. Palassini, Ann. Physik **18**, 868 (2009).
[11] A. Möbius, M. Richter, and B. Drittler, Phys. Rev. B **45**, 11568 (1992).
[12] Q. Li and P. Phillips, Phys. Rev. B **49**, 10269 (1994).
[13] A. Möbius, P. Karmann, and M. Schreiber, J. Phys.: Conf. Ser. **150**, 022057 (2009).
[14] A. Möbius and M. Richter, Phys. Rev. Lett. **105**, 039701 (2010).
[15] A. L. Efros, B. Skinner, B. I. Shklovskii, Phys. Rev. B **84** 064204 (2011).
[16] M. Goethe and M. Palassini, unpublished.
[17] S. D. Baranovskii, A. L. Efros, B. L. Gelmont, and B. I. Shklovskii, J. Phys. C **12**, 1023 (1979).
[18] A. Glatz, V. M. Vinokur, J. Bergli, M. Kirkengen, and Y. M. Galperin, J. Stat. Mech. **P06006** (2008).
[19] M. Goethe and M. Palassini, poster presented at this Conference.
[20] J.P. Sethna, in *Lecture Notes of the Les Houches Summer School 2006*, ed. J.-P. Bouchaud, M. Mézard, and J. Dalibard (Elsevier, Amsterdam, 2007), p. 257.
[21] O. Perkovic, K. A. Dahmen, and J. P. Sethna, Phys. Rev. Lett. **75**, 4528 (1995); K. Dahmen and J. P. Sethna, Phys. Rev. B **53**, 14872 (1996).
[22] F. Pázmándi, G. Zaránd, and G. T. Zimányi, Phys. Rev. Lett. **83**, 1034 (1999).
[23] P. Le Doussal, M. Müller, and K. Wiese, EPL **91**, 57004 (2010).
[24] A. A. Pastor, V. Dobrosavljevic, M. L. Horbach, Phys. Rev. B **66**, 014413 (2002).
[25] S. D. Baranovskii, B. I. Shklovskii, and A. L.Efros, Sov. Phys. JETP **60**, 1031 (1984)
[26] M. Lee, J.G. Massey, V.L. Nguyen, and B.I. Shklovskii, Phys. Rev. B **60**, 1582 (1999).
[27] D. Monroe, A. C. Gossard, J. H. English, B. Golding, and W. H. Haemmerle, Phys. Rev. Lett. **59**, 1148 (1987).
[28] J. Delahaye, J. Honoré, and T. Grenet, Phys. Rev. Lett. **106**, 186602 (2011).
[29] P. Le Doussal and K. J. Wiese, Phys. Rev. E **79**, 051106 (2009).
[30] T. Wappler, T. Vojta, and M. Schreiber, Phys. Rev. B **55**, 6272 (1997).

Journal of Physics: Conference Series **376** (2012) 012010 doi:10.1088/1742-6596/376/1/012010

Hopping Conductivity Enhanced by Microwave Radiation

Z. Ovadyahu

The Racah Institute of Physics
The Hebrew University, Jerusalem 91904, Israel

E-mail: zvi@vms.huji.ac.il

Abstract. Hopping conductivity is enhanced when exposed to microwave (MW) fields. Data taken on several Anderson-localized systems and granular-aluminium are presented to illustrate the generality of the phenomenon. It is suggested that the effect is due to a field-enhanced hopping, which is the ac version of a non-ohmic effect familiar from studies in the dc transport regime.

1. Introduction

Non-ohmic effects are commonly encountered in hopping conductivity. Deviations from linear transport conditions in these systems are actually difficult to completely avoid. Linear response conditions are expected when the applied field F obeys:

$$F \leq \frac{k_B T}{eL} \qquad (1)$$

Here, k_B is Boltzmann-constant; e is the electron charge, and T the temperature. L is the spatial scale over which the electron gains energy from the field F before it is dissipated into the bath, usually by phonon emission, thus $L=L(T)$. As the temperature gets lower L becomes larger, and maintaining Ohmic conditions by is possible only by reducing the potential drop across the sample. The need to ensure good signal-to-noise makes this harder at low temperatures.

Theoretical models to describe the modification to the conductance by a non-ohmic dc field predict an excess conductance ΔG that, at small fields, is exponential with eFL/k_BT.

The effect of electromagnetic waves on hopping conductivity has been given much less attention. Of particular interest is the case of microwave (MW) radiation where the wavelength λ is larger than a typical sample-size [4]. When exposed to MW fields, hopping conductance is enhanced which, perhaps naturally, was associated with heating [5]. Later studies, exploiting some unique properties of electron-glasses, suggested however that this is an adiabatic, non-equilibrium effect [6]. A characteristic feature of the results, which makes it difficult to reconcile the effect with simple heating, is the sub-linear dependence of ΔG on the microwaves power P.

In this paper we report on MW-excitation results obtained on films of different hopping systems; beryllium, granular-aluminum, In_2O_{3-x} (crystalline indium-oxide), and In_xO (amorphous indium-oxide). In all cases the ΔG at $T\approx$4K turn out to be sub-linear with the MW power in agreement with the results of previous studies. It is further shown that the functional dependence of $\Delta G(P)$ is consistent with the current-voltage characteristics measured independently (at low frequency) on the same sample. The MW-enhanced conductance is then conjectured to be a non-ohmic effect, of a similar nature as the field assisted-hopping phenomenon that is responsible to deviations from linear transport. The temperature dependence of the MW-induced ΔG is shown to be consistent with this conjecture.

2. Experimental

Several materials were used to prepare samples for measurements in this study. These were thin films of In_2O_{3-x}, In_xO, and granular-aluminum. Lateral size of the samples was typically 1mm, and thickness varied between 50 to 210 Å. Samples from these materials were similarly prepared by e-gun evaporation of In_2O_3 or aluminum pellets unto room temperature glass substrates in partial oxygen pressures of $4\text{-}6\cdot10^{-5}$ mbar and rates of 0.3-1 Å/second. The In_2O_{3-x} samples were obtained from the as-deposited amorphous films by crystallization at 525 K.

Conductivity of the samples was measured using a two terminal ac technique employing a 1211-ITHACO current preamplifier and a PAR-124A lock-in amplifier. The ac voltage bias was small enough to ensure near-ohmic conditions (except for the I-V measurements). Unless otherwise mentioned, measurements reported here were performed at T≈4.1K. Complementary details of sample preparation, characterization, and measurements techniques are given elsewhere [7].

The high-power synthesizer HP8360B, was used with power up to 25dBm (≈320 mW) for the MW excitation. The frequency range used in this work was 2-4 GHz. The output of the synthesizer was fed to the sample chamber via a coaxial cable ending with a short antenna.

3. Results and Discussion

The single most characteristic feature of the MW-enhancement phenomenon is the functional form of the excess conductance ΔG versus the radiation power P. In more than 120 samples studied in our laboratory ΔG increased with P less than linearly. Typical results of $\Delta G(P)$ scans, illustrating the sub-linear form of the response, are shown in figure 1:

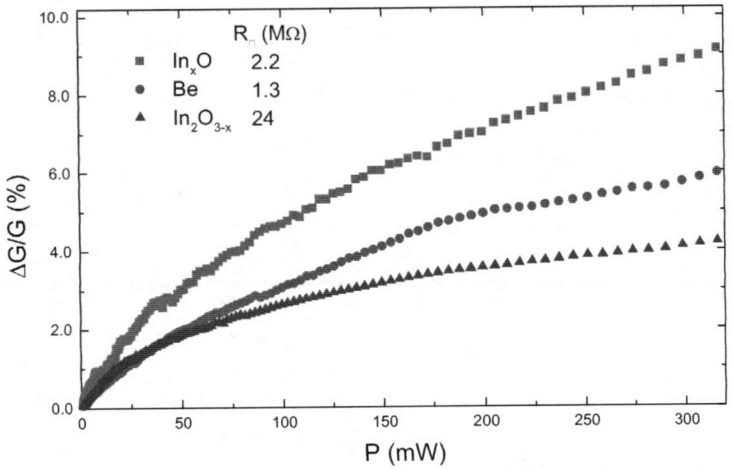

Figure 1: The fractional change of the conductance for three different electron-glasses as function of the MW power. Data taken at T=4.1 K

While we cannot rule out linearity at sufficiently small P, it is noted that $\Delta G(P)$ is sub-linear even for $\Delta G/G \leq 10\text{-}2$. This was the first indication that the origin of the effect is not likely to be 'heating' [6]. As more data became available, a correlation was found between the sensitivity of G to the voltage used in the conductance measurement, and the magnitude of ΔG produced by a given power of MW; in particular, samples that showed ohmic behavior up to high voltages, exhibited very small MW-induced ΔG. It was then natural to look for a correlation in the functional form of the two effects, which turned out to be the key feature in the search for the underlying mechanism.

Journal of Physics: Conference Series **376** (2012) 012010 doi:10.1088/1742-6596/376/1/012010

Figure 2 illustrates that $\Delta G(P^{1/2})$ is indeed similar to the $G(V)$ derived from the *I-V* characteristics of the sample. This similarity was observed in other samples. We are led therefore to consider the following heuristic picture: The MW radiation induces field across the sample (picked-up by the wires connected to the sample for the two-terminal conductance measurement). Once the potential difference associated with this field is greater than the voltage employed for measuring G, the conductance of the sample is modified in a similar vein as in the process of measuring G(V) directly (by dc or low frequency). In other words, the conductance monitored by a low-bias, low-frequency technique reflects the reduction of the hopping barriers induced by the high-bias associated with at the MW frequency. In this picture the reason for the sub-linearity of $\Delta G(P)$ is just the less-than-quadratic dependence of the excess conductance on an applied field. This mechanism is plausible when the wave-length λ is larger than the sample size which clearly is the case here ($\lambda \cong$10-30 cm for the MW frequencies used while sample size is $\cong$1 mm).

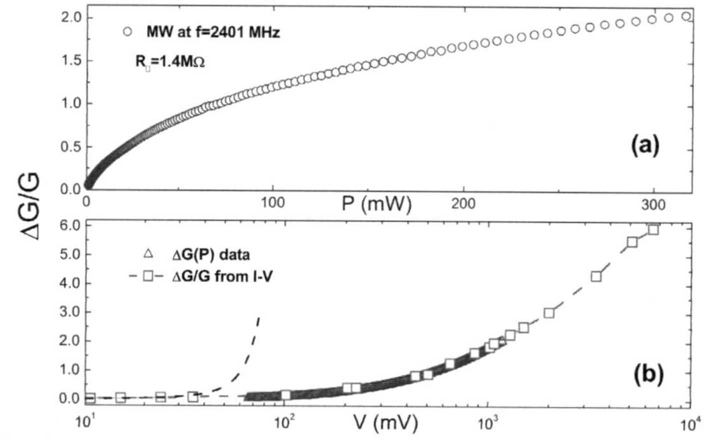

Figure 2: The fractional change of the conductance for (**a**) a typical granular-aluminium sample as function of the MW power, and (**b**) the same sample as function of the applied voltage (measured at f=73 Hz), and compared with $\Delta G(P^{1/2})/G$ based on the MW data by scaling the x-axis of $\Delta G(P^{1/2})$ by a constant factor of $\approx$ 65. The black dashed line depicts *exp(F)* dependence, which is shown to fits only a narrow range near the onset of non-ohmicity (see text).

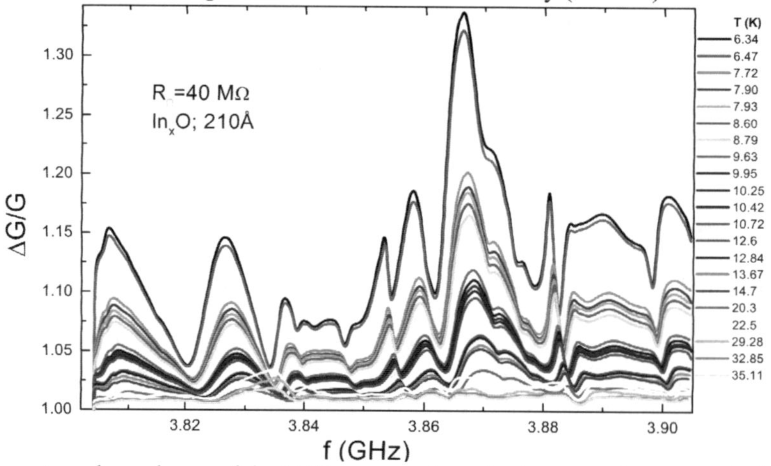

Figure 3: Temperature dependence of the MW enhanced conductance made on a typical amorphous indium-oxide film under a constant power of 25dBm. Note the shift of the $\Delta G(f)$ peak positions with temperature. This made it necessary to record a range of frequencies. The value for ΔG used in the plot of figure 4 was obtained by averaging over the range monitored here.

Journal of Physics: Conference Series **376** (2012) 012010 doi:10.1088/1742-6596/376/1/012010

Another set of data that are in line with the conjecture that the MW-enhanced conductance is a non-ohmic effect, is the temperature dependence of the excess conductance (at constant MW power). In these experiments, a range of frequencies was scanned at each temperature. This was done to cater for drifts in the cavity characteristics with the concomitant shift in conductance-response peaks (see figure 3 below). Figure 4 shows $\Delta G(T)$ due to the MW radiation normalized by $G(T)$ at zero field. This was obtained from the data of figure 3 in conjunction with the resistance versus temperature of this sample (inset to figure 4).

The conductance for fields that just take the system away from the ohmic regime is given [1,2,3] by $G=G_0 \cdot exp[eFL(T)/k_BT]$. For small $\Delta G/G$ this yields $\Delta G/G \propto eFL(T)/k_BT$. It has been suggested that $L(T)$ is the hopping length [3] or the percolation radius [2]. The fit to the data in figure 4 gives $\Delta G/G \propto T^{-1.81}$. A power-law dependence is actually in agreement with either of these theoretical expectation: considering the hopping-law exhibited by the sample in figure 4, and if $L(T)$ is the hopping length, one expects $\Delta G/G \propto T^{-1.5}$ while $\Delta G/G \propto T^{-2}$ if $L(T)$ is the percolation radius. It would be interesting to see a similar set of data on other systems that exhibits hopping conductivity as it might help to distinguish between the different theoretical approaches.

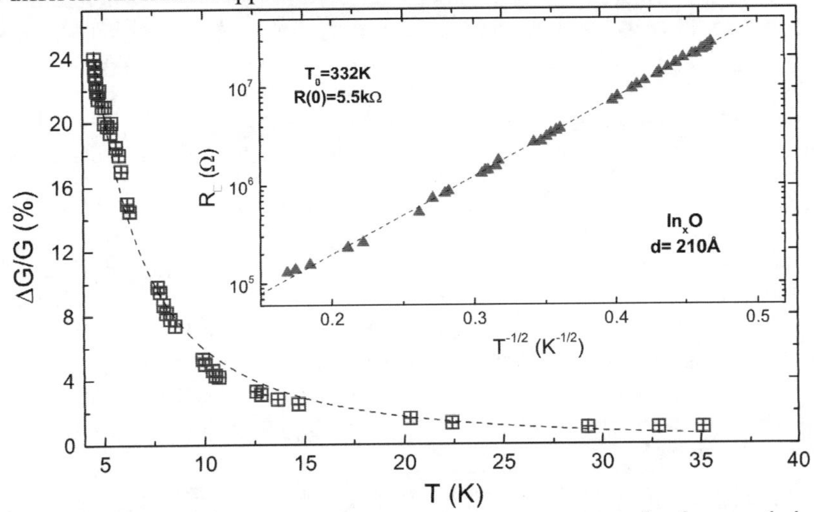

Figure 4: The fractional change of the conductance versus temperature for the sample in figure 4 (see text for details). The inset shows the temperature dependence of the sample resistance and some of its hopping parameters extracted on basis of $R_\square = R(0) \cdot exp(T_0/T)^{1/2}$.

While the temperature dependence is apparently not at variance with our picture, the magnitude of the effect appears to be problematic: By conjecture $\Delta G/G$ is a functional of $eFL(T)/k_BT$, and to obtain $\Delta G/G$ of order of 0.1-1 (which is the common range at the maximum power at 4 K), this functional has to be of order 1. For a typical field of 1-10 V/cm, it would appear that an unreasonably large L is required.

This problem is not specific to the MW effect; it is a problem of non-ohmicity in general. This, in fact, is a common occurrence in hopping conductivity: Values of L that are larger by 1-3 orders of magnitude than reasonably expected are reported in the literature [8]. These include results for samples doped by nuclear-transmutation [9] (presumed to be free of technical inhomogeneities). This is also the case in all our samples. For example, deviations from Ohm's law for the sample in figure 2 are already evident at $V \leq 50$ mV, which in terms of eq.1 yield $L \approx 10$ μm. This is clearly much too large a scale for a hopping sample at 4 K. It should be remarked that the origin of the discrepancy may not be that of L but rather a result of inhomogeneous field distribution. There seems to be some evidence for the latter in studies of non-ohmic behavior as function of sample-size [10].

Journal of Physics: Conference Series **376** (2012) 012010 doi:10.1088/1742-6596/376/1/012010

For demonstrating the similarity between *G(V)* and the MW-induced ΔG we have used the measured *G(V)* rather than the theoretical [1,2,3] expression,. These models, predicting $ln(G/G_0) \propto F$ (which incidentally, depict ΔG as sub-linear with *P*), might show agreement with our experiments (and indeed with other experiments) only over a very limited range. In [9] for example, a fit with the theory *may* be obtained only for $\Delta G/G \leq 0.08$, which is unconvincing for an exponential dependence. This is also the range over which this expression may fit our data while it progressively overestimates them for $\Delta G/G \geq 0.1$ (see figure 2b above). The model of Apsley and Hughes [11] which predicts $ln(G/G_0) \propto F^2$ may be valid near the very onset for non-ohmicity (where, for symmetry reasons, the quadratic dependence is expected), but it is even further from the experimental dependence over the relevant range. It is noted that the lack of agreement with theory is not related to the use of ac technique in the experiments (by Grannan et al [9] and in our experiment) while the theories refer to a dc measurement; It turns out that the functional form of *G(V)* measured by dc is essentially similar to that measured by a low frequency ac technique [12].

There are two additional issues that should be clarified. The first concerns the mapping of $\Delta G(P^{1/2})$ onto *G(V)* by rescaling the voltage axis. The procedure, exemplified above in figure 2, was tried on many samples with similar success. However this procedure never gave a perfect registry of the two curves. However, it is not clear that the two measurements should fit better than they do; The *G(V)* measured at different frequencies will, in general, be different. The reason is that the Miller-Abrahams resistors that make up the hopping system are actually impedances with value that depend on frequency. As a result, the local potential drops across the sample will in general be different at different frequencies.

The second issue is more subtle. Applying a low-frequency, non-ohmic field across any of the samples described here does more than just increase their conductance. When the field is allowed to act on the sample for a certain time the conductance is not stationary; it slowly increases with time, and after the field is set back to achieve linear-response, the conductance does not immediately resume its original equilibrium value. Rather, it is initially higher by a certain ΔG, which slowly relaxes to zero [13]. By contrast, when the MW is applied with amplitude such that the resulting ΔG is a large as achieved by the respective low-frequency field, the conductance does not show these glassy features [6]. The lack of these non-equilibrium features when the sample is "stressed" by MW is illustrated in figure 5 for three different materials:

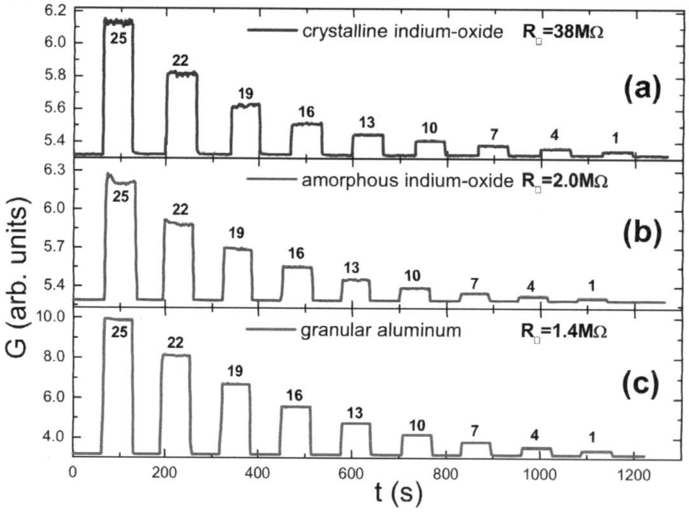

Figure 5: Response of three samples to MW exposure of different power (marked in dBm over time segments when MW is turned on). Note that, after the initial jump, the conductance does not keep on increasing while the MW is on, and it promptly resumes its 'equilibrium' value when the MW is

Journal of Physics: Conference Series **376** (2012) 012010 doi:10.1088/1742-6596/376/1/012010

turned off even at the highest power. Data taken at T=4.1 K, at f= 2.530, 2.241, and 2.401 GHz for (a), (b), and (c) respectively.

In contrast with the similarity in terms of enhancing the hopping conductance, there is evidently a fundamental difference between the MW and the low-frequency field mechanisms. This difference is presumably associated with the ability of the electronic system to absorb energy from the field; the slow relaxation of ΔG that follows exposure to the field is the telling sign of stored excess *energy*. The latter is accumulated during the time the field is on and this is indicated by the slow build-up of excess conductance. Note that, by itself, a conductance increase does not necessarily imply energy absorption as the process may be adiabatic. Specific experiments to elucidate this issue by using fields with different frequencies will be presented elsewhere.

Acknowledgments

Illuminating discussions with A. Frydman and A. Goldman are gratefully acknowledged. I am indebted to M. Pollak and B. Shapiro for useful comments and critical remarks. This work was supported by a grant administered by the US-Israel Science Foundation and by the Israeli Academy of Sciences.

References

[1] Hill R M, 1971, *Phil. Mag.* **24** 1307
[2] Shklovskii B I, 1976 *Fiz. Tech. Polouprovodn.* **10** 1440 [*Sov. Phys. Semicond.* **10** 855]
[3] Pollak M and Riess I, 1976 *J. Phys. C* **9** 2339
[4] Zvyagin I P, 1978 *Phys. Stat. Solidi* 88 149
[5] Ben-Chorin M, Ovadyahu Z, and Pollak M, 1993 *Phys. Rev. B* **48** 15025
[6] Ovadyahu Z, *Phys. Rev. Lett.* 2009 **102** 206601
[7] Vaknin A, Ovadyahu Z, and Pollak M, 2002 *Phys. Rev. B* **65**, 134208
[8] Rosenbaum T F, Andres K, and Thomas G A, 1980 *Solid State Commun.* **35** 663
Faran O, and Ovadyahu Z, 1988 *Solid State Comm.*, **67**, 823
[9] Grannan S M, Lang A E, Haller E E, and Beeman J W, 1992 *Phys. Rev.* B **45** 4516
[10] Kowal D and Ovadyhau Z, 2008 *Physica C*, **468**, 322
[11] Apsley N and Hughes H P, 1974 *Philos. Mag.* **30**, 963
[12] Ovadyahu Z (unpublished)
[13] Orlyanchik V, Vaknin A, and Ovadyahu Z, 2002 *Phys. Stat. Sol.*, **b230**, 67
Orlyanchik V, and Ovadyahu Z, 2004 *Phys. Rev. Lett.*, **92**, 066801

Journal of Physics: Conference Series **376** (2012) 012011 doi:10.1088/1742-6596/376/1/012011

Electric field dependence of charge-carrier hopping transport at large carrier concentrations in disordered organic solids: Meyer-Neldel and Gill energies

I I Fishchuk[1], A Kadashchuk[2], Mujeeb Ullah[3], H Sitter[3], N S Sariciftci[4] and H Bässler[5]

[1] Institute for Nuclear Research, NASU, Prospect Nauky 47, 03680 Kyiv, Ukraine
[2] Institute of Physics, NASU, Prospect Nauky 46, 03028 Kyiv, Ukraine
[3] Institute of Semiconductor & Solid State Physics, Johannes Kepler University of Linz, Austria
[4] Linz Instute of Organic Solar Cells, Johannes Kepler University of Linz, Austria
[5] Universität Bayreuth, Lehrstuhl Experimentalphysik II, Bayreuth, Germany

E-mail: ifishch@kinr.kiev.ua

Abstract. Effective medium approach has been extended to describe the temperature dependent hopping charge-carrier mobility at arbitrary electric fields in the large carrier density transport regime. We take into account the spatial energy correlations in organic materials with Gaussian disorder. The theory is applied to describe recent experimental measurements of the electron transport properties in a C_{60}-based OFET for different lateral electric fields F_{DS}. Since this model is not limited to zero-field mobility, it allows a more accurate evaluation of important material parameters from experimental data measured at a given electric field. The shift of the Meyer-Neldel energy E_{MN} upon applied lateral electric field F_{DS} and the Gill energy E_G upon the gate voltage V_G in an OFET is shown to be a consequence of the spatial energy correlation effects in the organic semiconductor film. We showed that both the Meyer-Neldel and Gill energies can be used for estimating the width of the Gaussian density-of-states distribution.

1. Introduction

Charge carrier transport in organic materials has been consistently described by a formalism based on hopping in a Gaussian density-of-states (DOS) distribution. The Gaussian Disorder model (GDM) originally suggested by Bässler and co-authors [1], has been developed specifically for the low-carrier-density hopping transport regime. For large carrier concentrations the charge carrier mobility depends also on carrier concentration which has been recently described within an Extended Gaussian Disorder model (EGDM) [2-5]. Despite the success of the disorder formalism in rationalizing many aspects of the charge transport in organic semiconductors, the electrical transport mechanisms in realistic organic electronic devices are still not completely understood. This applies, for instance, to the dependence of the charge carrier mobility on the electric field (F) and, in particular, to the $\ln \mu \propto F^{1/2}$ "Poole-Frenkel" (PF) type dependence being experimentally observed in a rather broad field range. Parris et

[1] To whom any correspondence should be addressed

al. [6] and Novikov et al. [7] have suggested on the basis of their computer simulations of the charge transport in the low-carrier-concentration limit with accounting for long-range spatial energy correlations a so-called Correlated Disorder model (CDM). The description of charge transport at high carrier concentrations was addressed just recently using numerical simulations Novikov [8] and Bouhassoune et al. [9] suggested a so-called Extended Correlated Disorder model (ECDM).

In the present study we suggest an analytical Effective medium approach to describe charge mobility in disordered organic solids, which is formulated for arbitrary electric fields at high carrier concentrations and takes into account spatial energy correlations. We applied this theory to analyze recent experimental data on electron mobilities measured in a benchmark OFET device based on C_{60} semiconductor films. We study the influence of the electric field on the so-called Meyer-Neldel (MN) compensation rule [10] typically observed for the temperature dependence of the charge carrier mobility in a high-carrier-concentration regime. In context of the charge transport in OFETs, the MN rule suggests a empirical relation between the Arrhenius activation energy E_a and the mobility prefactor. As our previous treatment [11] was limited to zero-electric field, we now extended it for arbitrary electric fields to study simultaneously the effect of the carrier concentration and the electric field on the Meyer-Neldel phenomenon for the OFET mobility. The presented theory was also used to describe the recently observed MN-type effect in OFETs regarding $\mu(T)$ dependence upon varying the electric field.

2. Theoretical formulation

Our analytic theoretical treatment is focused on the description of the electric-field dependence of the drift charge-carrier mobility in a random organic system at different temperatures and carrier concentrations, and is based on an Effective Medium approximation (EMA). Let us consider a random 3D hopping transport system with an applied electric field $\mathbf{F} = \{F, 0, 0\}$. In general the effective drift hopping mobility μ_e can be obtained as

$$\mu_e = a k_0 \frac{W_e^+ - W_e^-}{F}, \tag{1}$$

where $W_e^\pm = \langle \tau_{12}^\pm \rangle^{-1}$, $\langle \tau_{12}^\pm \rangle = \int_{-\infty}^{\varepsilon_t} P(\varepsilon) \{W_{12}^\pm(\varepsilon_t, \varepsilon)\}^{-1} \mathrm{d}\varepsilon \Big/ \int_{-\infty}^{\varepsilon_t} P(\varepsilon) \mathrm{d}\varepsilon$.

Coefficient k_0 emerges in Eq. (1) to include the generalized Einstein equation. We use a Miller-Abrahams-type jump rate $W_{12}^\pm(\varepsilon_t, \varepsilon) = W_0 \exp\{-[(|\varepsilon_t - \varepsilon \mp eaF| + (\varepsilon_t - \varepsilon \mp eaF))/2k_B T]\}$. Here $W_0 = \nu_0 \exp(-2r_t/b)$, where ν_0 is the attempt-to-escape frequency, r_t is the jump distance below the effective transport energy ε_t [12], $a = N^{-1/3}$, N is the site concentration and b is the localization radius of the charged site. Configuration averaging $\langle ... \rangle$ of the hopping transitions times has to be done over the energy distribution of empty localized states, viz. by using the function $P(\varepsilon) = g(\varepsilon)[1 - f(\varepsilon, \varepsilon_F)]$, where $f(\varepsilon, \varepsilon_F)$ is the Fermi-Dirac distribution. The Fermi energy level ε_F can be determined from the following transcendental equation $n = \int_{-\infty}^{\infty} d\varepsilon\, g(\varepsilon) f(\varepsilon, \varepsilon_F)$ for the carrier concentration n. The coefficient k_0 (cf. Eq. (1)) in this case can be determined as $k_0 = 1 - \int_{-\infty}^{\infty} d\varepsilon\, g(\varepsilon) f^2(\varepsilon, \varepsilon_F) \Big/ \int_{-\infty}^{\infty} d\varepsilon\, g(\varepsilon) f(\varepsilon, \varepsilon_F)$. In the present study we assume a Gaussian DOS distribution $g(\varepsilon)$ with the width σ as generally accepted to be appropriate for disordered

Journal of Physics: Conference Series **376** (2012) 012011 doi:10.1088/1742-6596/376/1/012011

organic media $g(\varepsilon)=\left(N/\sigma\sqrt{2\pi}\right)\exp\left[-(1/2)(\varepsilon/\sigma)^2\right]$. After combining, we can express the effective charge carrier mobility μ_e as

$$\mu_e=\mu_0 k_0\left(\frac{r_t}{a}\right)^2\exp\left(-2\frac{r_t}{b}\right)\frac{\left(Y_e^+\right)^{-1}-\left(Y_e^-\right)^{-1}}{f}, \tag{2}$$

where

$$Y_e^{\pm}=\frac{\int_{-\infty}^{x_t}dt\frac{\exp\left\{-\frac{t^2}{2}+\frac{1}{2}\left[\left|x_t-t\mp f\right|+\left(x_t-t\mp f\right)x\right]\right\}}{1+\exp\left[-\left(t-x_F\right)x\right]}}{\int_{-\infty}^{x_t}dt\frac{\exp\left\{-\frac{t^2}{2}\right\}}{1+\exp\left[-\left(t-x_F\right)x\right]}}. \tag{3}$$

Here $Y_e^{\pm}=W_e^{\pm}/W_0$, $f=eaF/\sigma$, $x=\sigma/k_BT$, $x_F=\varepsilon_F/\sigma$, $x_t=\varepsilon_t/\sigma$, $\mu_0=ea^2\nu_0/\sigma$. The effective transport energy ε_t in our method does not depend on the applied electric field F and hence for a Gaussian DOS distribution can be determined from the following transcendental equation derived for zero-field mobility [3]

$$\frac{1}{\sqrt{2\pi}}\frac{\exp\left(-\frac{1}{2}x_t^2\right)}{1+\exp\left[-\left(x_t-x_F\right)x\right]}\left\{\frac{1}{\sqrt{2\pi}}\int_{-\infty}^{x_t}dt\frac{\exp\left(-\frac{1}{2}t^2\right)}{1+\exp\left[\left(t-x_F\right)x\right]}\right\}^{-\frac{4}{3}}=\frac{3}{2}\left(\frac{4\pi}{3B}\right)^{\frac{1}{3}}x\frac{b}{a}. \tag{4}$$

Here parameter $B=2.7$ being determined according to percolation criteria [13]. Factor r_t is calculated by

$$r_t=a\left\{\frac{4\pi}{3B}\frac{1}{\sqrt{2\pi}}\int_{-\infty}^{x_t}dt\frac{\exp\left(-\frac{1}{2}t^2\right)}{1+\exp\left[-\left(t-x_F\right)x\right]}\right\}^{-\frac{1}{3}}. \tag{5}$$

The results of our analytical calculation by Eq. (2) are in very good agreement with the computer simulations data by Pasveer et al. [2] obtained within the EGDM approximation for temperature-, field- and carrier-concentration- dependences of the charge carrier mobility in a random (non-correlated) organic system. It is well established that a PF-type field dependence of the charge carrier mobility can be reproduced over an extended interval of electric fields only when some kind of correlated disorder is taken into account in the framework of the Correlated Disorder model (CDM) or the Extended Correlated Disorder model (ECDM).

Hence, hereafter in our theoretical description we will use the concept of energy correlations. Generally, accounting for the energy correlated disorder effects is not amenable to analytical treatment, although it might be solved by computer simulations [8, 9].

We therefore will use here in our analytical treatment some key ingredients from the numerical simulations studies performed before within the CDM and ECDM [7, 8, 9], in particular, (i) the width of the so-called correlated DOS, σ_c, has to be somewhat smaller than the initial non-correlated σ, (ii) the typical jump length, governing the effective charge carrier mobility, depends not only on electric field but also on carrier concentration. Thus, for calculation of the field-, temperature-, and carrier-concentration- dependences of the charge carrier mobility in disordered organic semiconductors we will hereafter use Eq. (2-5) obtained within EGDM approximation with accounting for energy correlation effects by substituting the parameters σ, f by σ_c, $f_c y$, respectively, where $\sigma_c = 0.83\sigma$, $y = (1 - n/N)^m$, $m = -9.69 + 10.81(\sigma/k_B T) - 0.905(\sigma/k_B T)^2$, $f_c = h_c\sqrt{x_c/2}$, $h_c = \sqrt{eaF/\sigma_c}$, $x_c = \sigma_c/k_B T$.

3. Effect of electric field and carrier concentration on "Meyer-Neldel" and "Gill" energies

We discriminate between the isokinetic temperatures resulting from intersection of $\ln(\mu) \propto T^{-1}$ upon varying carrier concentration and that upon varying applied electric field and call them Meyer-Neldel- and Gill- temperature, respectively. According to the results, the Meyer-Neldel temperature (denoted as T_1) and Gill temperature (denoted as T_2) are not constant but depend on the electric field F and relative carrier concentration n/N, respectively. This effect indeed can be well reproduced within the present EMA model. The field dependence of T_1 can be parameterized as follows

$$\frac{k_B T_1}{\sigma} = 0.5 + 0.029\frac{eaqF}{\sigma} - 0.039\left(\frac{eaqF}{\sigma}\right)^2. \tag{6}$$

Here q is a parameter to be described below.

In line to that, the Gill-temperature T_2 tends to decrease with increasing carrier concentration n/N. This carrier density dependence of T_2 can be well approximated by

$$\frac{k_B T_2}{\sigma} = 0.416 - 0.037\log_{10}\left(\frac{n}{N}\right) - 0.0031\left[\log_{10}\left(\frac{n}{N}\right)\right]^2. \tag{7}$$

From our analyze we conclude that the change of the MN-temperature T_1 upon applied electric field and the decrease of the Gill-temperature T_2 with increasing n/N results from the presence of energy correlation effects, namely, due to the decrease of the typical jump length with increasing carrier concentration. It should be pointed out that the above calculated dependences for T_1 and T_2 are only relevant to the range of electric fields where the PF-type field dependence holds.

The results of the above calculated charge carrier mobility in the high-carrier-concentration limit can be used to estimate the energetic disorder parameter σ from experimental data basically by two different methods: First, the experimentally measured MN-temperature T_1 ($E_{MN} = k_B T_1$) at a given electric field F (within the field interval where a PF-type dependence is obeyed) is inserted into Eq. (6). It results in a quadratic equation to calculate the parameter σ. The solution reads

$$\sigma = E_{MN} A\left(1+\sqrt{1+\left(\frac{eaqF}{E_{MN}}\right)^2 \frac{0.078}{A^2}}\right), \quad A = 1-0.029\frac{eaqF}{E_{MN}}. \tag{8}$$

The experimentally measured Gill-temperature T_2 ($E_G = k_B T_2$) at a given high carrier concentration n/N (being determined by a gate voltage (V_G) in an OFET) is substituted into Eq. (7), and yields the following relation for calculating σ.

$$\sigma = \frac{E_G}{0.416-0.037\log_{10}\left(\frac{n}{N}\right)-0.0031\left[\log_{10}\left(\frac{n}{N}\right)\right]^2}. \tag{9}$$

To use Eq. (9) one has, however, to know the effective carrier concentration in a thin conductive channel of an OFET, which demonstrates a highly non-uniform distribution - strongly decreasing from the semiconductor/insulator interface into the bulk. Thus, the first of the above-mentioned methods seems to be more appropriate for estimating the disorder parameter σ from experimental data. A big advantage of the present theoretical model is that it does not require an extrapolation of experimental data to zero electric field.

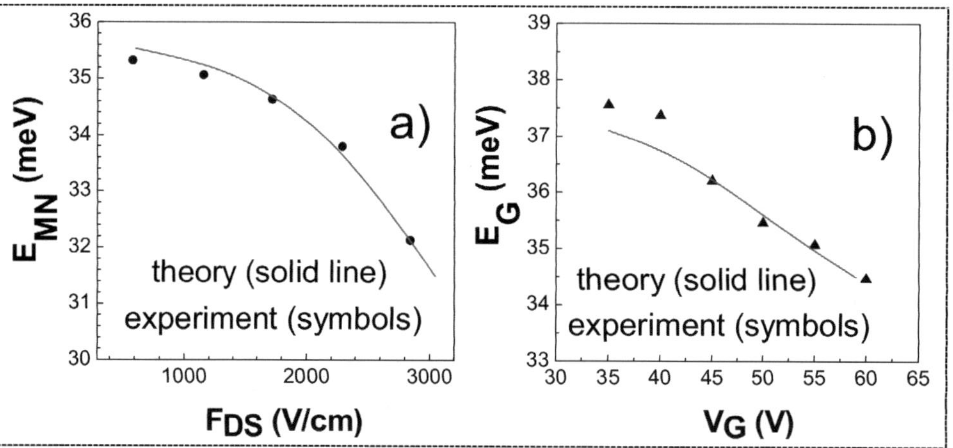

Figure 1. Experimentally determined (a) MN-energy $E_{MN} = k_B T_1$ (symbols) versus the applied lateral electric field and (b) Gill-energy $E_G = k_B T_2$ as a function of used gate voltage V_G. Fitting by Eqs. (2-5) using the following set of parameters: $\sigma = 0.07$ eV; $a/b = 5$; a=1.4 nm and $q = 256$ are given by solid.

For quantitative description of the lateral-field dependence of the OFET mobility one has to involve a concept of a strong inhomogeneity of the lateral electric field inside the accumulation layer formed in an OFET caused by an inhomogeneous morphology of the semiconductor, which results in strong local fields confined to specific places establishing major energetic barriers and, hence, controlling the overall (effective) hopping charge mobility through the OFET channel, as suggested elsewhere [14]. Thus, it is actually the effective local field that should be used in Eqs. (6, 8) instead of the average applied electric field. The ratio q between the local electric field strength and that averaged over the transistor channel V_d / L can be determined from the field dependence of E_{MN}.

In fact, a set of just two E_{MN} values measured at two different electric fields inserted into Eq. (8) enables calculating σ and q parameters. The effects observed in Figure 1(a) and (b) can only be described provided that energy correlation effects are taken into account. This suggests an important experimental test for the presence of energy correlations in an organic semiconductor.

4. Conclusion

An analytical EMA theory was formulated to describe the effect of electric field on the charge carrier mobility at high carrier concentrations in disordered organic semiconductors at large carrier concentrations with accounting for energy correlation effects. The EMA theory predicts that the Meyer-Neldel temperature (T_1) and Gill temperature (T_2) are not constant but depend on the electric field and carrier concentration, respectively; and this found to be a consequence of spatial energy correlations in organic disordered materials. We show that both above temperatures can be used for estimating the energy disorder parameter – the width of the DOS distribution. The present model allows more accurate evaluation of important material parameters from experimental data measured at arbitrary electric field without extrapolation of experimental data to the zero-electric field.

Acknowledgements
The research was supported by the Austrian Science Foundation, projects S9706 and S9711 and the Science & Technology Center in Ukraine under the contract No. 5258, the ÖAD Project UA 10/2011.

[1] Bässler H 1993 *Phys. Stat. Sol. (b)* **175** 15
[2] Pasveer W F, Cottaar J, Tanase C, Coehoorn R, Bobbert P A, Blom P W M, de Leeuw D M and Michels M A J 2005 *Phys. Rev. Lett.* **94** 206601
[3] Fishchuk I I, Arkhipov V I, Kadachchuk A, Heremans P and Bässler H 2007 *Phys. Rev.* B. **76** 045210
[4] Craciun N I, Wildeman J and Blom P W M 2008 *Phys. Rev. Lett.* **100** 056601 (2008).
[5] Coehoorn R, Pasveer W F, Bobbert P A and Michels M A J 2005 *Phys. Rev.* B **72** 155206
[6] Parris P E, Dunlap D H and Kenkre V M 2000 *Phys. Stat. Sol.* (b) **218** 47
[7] Novikov S V, Dunlap D H, Kenkre V M, Parris P E and Vannikov AV 1998 *Phys. Rev. Lett.* **81** 4472
[8] Novikov S V 2008 *Phys. Stat. Sol.* (c) **5** 740
[9] Bouhassoune M, van Mensfoort M S L, Bobbert P A and Coehoorn R 2009 *Organic Electronics* **10** 437
[10] Meyer W and Neldel H, 1937 *Z. Tech. Phys.* **18** 588
[11] Fishchuk I I, Kadashchuk A K, Genoe J, Mujeeb Ullah, Sitter H, Singh Th B, Sariciftci N S, and Bässler H 2010 *Phys. Rev.* B **81** 045202
[12] Baranovskii S D, Cordes H, Hensel F, and Leising G 2000 *Phys. Rev.* B **62** 7934
[13] Rubel O, Baranovskii S D, Thomas P and Yamasaki S 2008 *Phys. Rev.* B **69** 014206
[14] Fishchuk I I, Kadachchuk A, Ullah Mujeeb, Sitter H, Pivrikas A, Genoe J, and Bässler H 2011 *Phys. Rev. B, (submitted).*

Journal of Physics: Conference Series **376** (2012) 012012 doi:10.1088/1742-6596/376/1/012012

Extension of Anderson's Localization Theory to Interacting Systems.

M. Pollak

Dept. of Physics, University of California, Riverside, CA 92502

pollak@ucr.edu

Abstract. The paper adapts Anderson's original paper on localization to include interaction. For that purpose his perturbation expansion is extended to a many-body Rayleigh Schrödinger perturbation expansion. Two systems are compared which are similar except that one does and the other does not include Coulomb interactions. The latter is assumed to have all states localized. It is shown that interactions have two opposing effects – one enhancing localization, the other delocalization. Deep in the Anderson localization regime the first effect prevails, close to the Anderson transition the latter prevails

1. Introduction

Anderson's original paper [1] argued on the basis of a perturbation expansion that under large enough disorder conditions a particle could not propagate over a macroscopic distance. The criterion for localization in this approach is the convergence of the perturbation series. Such a convergence indicates that as higher order terms add local functions at more remote sites to the eigenfunction their amplitude keeps decreasing. In contrast, a divergent series indicates that amplitudes at very remote sites are still important, implying a delocalized wave function.

Since Coulomb interactions are important in most disordered systems their effect on Anderson localization has been of interest for some decades. Early arguments held that interactions in disordered systems are a source of additional disorder energy so they should enhance localization. Such arguments are basically Hartree type. An extension to Hartree-Fock [2] also found that interactions enhance localization. Neither of these include correlation effects and it was argued [3] that these act in favor of enhancing delocalization. More recent works on the subject found that short range interactions preserve localization [4] but long range interactions can induce delocalization [5,6]

This paper adapts [1] to include many-body effects, using a many-body Rayleigh- Schrödinger (RS) perturbation expansion which allows for studying effects of interaction. The paper takes of on previous work [7]. To evaluate the effect of interaction on localization the perturbation series are compared for an interacting and a non-interacting system.

As in [1], the basis set is taken to be the set of N local functions located around N sites. Slater determinants formed of these constitute the many-particle basis set into which the perturbation series is expanded. Interaction increases both the numerators and the denominators in the expansion thus indicating two diverse effects of interactions on localization. Arguments based on RS perturbation theory are presented that close to the Anderson transition interactions can cause delocalization where a similar non-interacting system is still fully localized. The said delocalization is of a collective nature.

Journal of Physics: Conference Series **376** (2012) 012012 doi:10.1088/1742-6596/376/1/012012

2. Physical Considerations

The one-particle RS expansion is stated in (1). For the purpose of this paper, the set of functions $|m_0\rangle$ are local functions, say 1s functions centered on sites m.

$$|i\rangle = |i_0\rangle + \sum_j \frac{\langle i_0|V|j_0\rangle}{\varepsilon_i - \varepsilon_j}|j_0\rangle + \sum_{j,k} \frac{\langle i_0|V|j_0\rangle}{\varepsilon_i - \varepsilon_j}\frac{\langle j_0|V|k_0\rangle}{\varepsilon_i - \varepsilon_k}|k_0\rangle + \sum_{j,k,l} \frac{\langle i_0|V|j_0\rangle}{\varepsilon_i - \varepsilon_j}\frac{\langle j_0|V|k_0\rangle}{\varepsilon_i - \varepsilon_k}\frac{\langle k_0|V|l_0\rangle}{\varepsilon_i - \varepsilon_l}|l_0\rangle + \ldots \tag{1}$$

$\langle m_0|V|n_0\rangle = t_{mn} \approx r_{mn}^{-1}\exp(-r_{mn}/\xi)$ are the transfer energies between sites m, n, ξ is the localization length, r_{mn} is the distance between sites m, n and ε_m is the random energy (distributed over an energy range W) of site m. For simplicity of notation the charge of the electron and the dielectric constant were taken to have the values 1. In the one-electron localization theory the important parameter is t/W where t is taken to be the value corresponding to the average nearest neighbour distance. The critical value of t/W where the Anderson transition occurs in 3D is not known precisely but is close to 1/20.The equation differs from the original Anderson perturbation expansion in that his expansion was for the time-dependent Schrödinger equation while (1) is for the stationary Schrödinger equation. Since at question is the nature of the eigenfunction (localized or delocalized), the latter is more convenient here. The expansion of (1) can be visualized as shown in figure 1. The dots represent sites and the lines of different colours represent successive terms in the perturbation series. When the successive terms become sufficiently smaller than their preceding terms the series converges. The wave function centred on i then cannot extend to infinity except with infinitesimal amplitude, so the eigenstate is localized. In the opposite case the series diverges, the amplitude at infinity is comparable to the amplitude at i and the particle is delocalized. As is well known, the possibility of forming loops puts a caveat on the former statement. However this appears to only affect the point at which the divergence occurs, not the divergence per se.

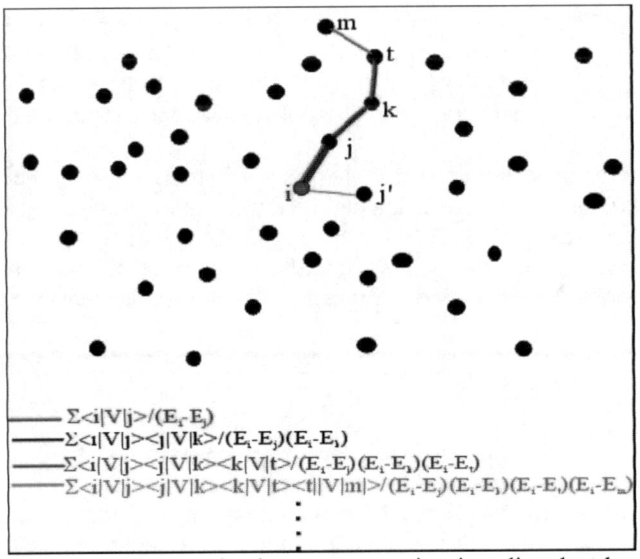

Figure 1. Aid to visualize equation 1. The dots represent sites in a disordered system. The lines represent the successive terms in the equation as indicated in the figure. Each term adds a local function with amplitude corresponding to the value of the term to the eigenfunction. The higher the order of the term the further the eigenfunction can spread.

To be able to deal with interactions the expansion is written below in a form to accommodate many-electron wave-functions.

$$|I\rangle = |I_0\rangle + \sum_J \frac{\langle I_0|V|J_0\rangle}{E_I - E_J}|J_0\rangle + \sum_{J,K} \frac{\langle I_0|V|J_0\rangle}{E_I - E_J}\frac{\langle J_0|V|K_0\rangle}{E_I - E_K}|K_0\rangle + \sum_{J,K,L} \frac{\langle I_0|V|J_0\rangle}{E_I - E_J}\frac{\langle J_0|V|K_0\rangle}{E_I - E_K}\frac{\langle K_0|V|L_0\rangle}{E_I - E_L}|L_0\rangle + \dots \quad (2)$$

(2) $|X_0\rangle$ (and the dots in the figure 1) are now Slater determinants (configurations) so the space in the figure becomes the phase space (or Fock space).

Both the interacting and the non-interacting case can be expressed in this form and compared with each other. For that purpose, the basis set is the same in both cases, namely the set of local functions as described above. The many-body basis set is the set of all Slater determinants formed of these local states.

To compare "apples with apples" one needs to compare similar situations, except for interactions, in both cases. In particular both systems should have the same ratio of electrons to sites N, say $n/N=½$. The interactions are strongest for such a ratio - for a smaller ratio the density of particles is smaller, for a larger ratio one can consider the holes as the interacting particles. In order to deal with a neutral system a positive charge $½|e|$ is placed on each site as is customary. To see whether interactions can enhance delocalization, assume that without interactions all states are localized and examine whether interactions can cause the RS series to diverge.

There is a formal problem with the expansion (2). In the thermodynamic limit the number of sites N goes to infinity so already the first sum in (2) has an infinite number of terms, each of which may be finite. The series therefore may diverge even when the numerators are very small. But the meaning of it merely is that $|I_0\rangle$ is connected to an infinite number of Slater determinants. One could argue that the system is delocalized in Fock space. But the question of interest is whether the system is delocalized in real space. In real space the first sum in (2), even if infinite, merely signifies that the wave function extends from each of the N sites to only nearby sites in real space. In that sense the inclusion of the term accounts only for an extension of the wave function through a short distance. Whether it extends over a microscopic distance is determined by the magnitude of the high order terms. The reasonable approach thus is to begin with a single site in $|I_0\rangle$ and examine the probability P that the amplitude of the high order terms connected to that site through all the lower order terms is comparable to the amplitude on this site. Since there are N sites in $|I_0\rangle$ the wave function will be delocalized in space if $P \geq 1/n$.

Since it is known that in three dimensions there is a transition in the absence of interactions, we need not actually execute such an expansion but merely see whether, in comparison with the non-interacting system the expansion for the interacting system enhances or diminishes P .

As will be shown below, interactions introduce two effects, one favoring enhancement of localization, the other enhancement of delocalization. The former involves the denominators, the latter the numerators.

Note at this point that in the case of Coulomb interactions there is a relation between t and E_C, defined as the Coulomb energy over an average nearest neighbor distance $\langle r\rangle$:

$$t=\langle r\rangle^{-1}\exp(-\langle r\rangle/\xi)\equiv E_C\exp(-\langle r\rangle/\xi) \quad (3)$$

with $t_c \approx W/20$ (in 3 dimensions) one has near the Anderson transition $W \approx E_C$. The importance of this relation will become clear later on.

Starting with the denominators, without interactions when I,J differ only in the occupation of two sites (*i.e.* an electron occupying a different site in I and in J) E_I-E_J is just the difference in the random energies ε_i-ε_j . i,j are, respectively, the two sites in I, J, that differ in occupation. In the presence of interactions the site energies include also interactions with all the other sites. Considering such interaction as an additional random energy (due to random spatial distribution), the effect of interactions on the denominators in (2) is to increase them - the spread of E_k is larger than the spread of the random energies ε_k. (It turns out that E_I-E_J is determined by the energies E_k of the sites k that differ in I and in J.) This basically corresponds to the old argument that interactions enhance localization. It should be remarked in passing that the definition of E_k here differs somewhat from the definition of Efros and Shklovskii [8]. Although in their definition and in the present definition E_k includes both the random and the interaction energies, in [8] E_k for an unoccupied site relates to an

n+1 electron system but in the present definition it always refers to an n-electron system, whether k is occupied or not.

Turning now to the effect of interaction on the numerators, without interaction V is a 1-particle operator so $\langle X_0|V|Y_0\rangle$ is non-zero only for Y_0, X_0 differing in the occupation of two sites. For a two-electron transition (involving four sites) $\langle X_0|V|Y_0\rangle$ includes terms with coefficients of the form

$$\int\alpha(x_i)\,\beta(x_j)\,V(x_i)\,\gamma(x_i)\,\delta(x_j)\,d(x_i)\,d(x_j) = \langle\alpha|V|\gamma\rangle\langle\beta|\delta\rangle = 0 \qquad (4)$$

where α (say) is the local function centered on site α. (3) vanishes because $\langle\beta|\delta\rangle$ is a dot product of two orthogonal functions. The integral of the product of all other electron coordinates in $\langle X_0|V|Y_0\rangle$ is $\prod_\upsilon\langle\upsilon|\upsilon\rangle = 1$. All exchange terms vanish, again because of orthogonality.

In the presence of interactions V is a two-particle operator so the terms corresponding to (4) involving $V(x_i,x_j)$ do not vanish: $\int\alpha(x_i)\,\beta(x_j)\,V(x_i,x_j)\,\gamma(x_i)\,\delta(x_j)\,d(x_i)d(x_j) \neq 0$. Thus, interactions add a large number of terms to the perturbation expansion (2). Furthermore, the Coulomb interaction between near neighbors amplifies this effect: due to electron-electron repulsion near neighbor pairs are likely to be singly occupied. This is particularly so near the Anderson transition where the disorder energy $W \approx E_C$.

Figure 2 illustrates the point: there are about twice as many possible transitions from k in K to some l in L with than without interactions. Furthermore, each such transition can be either a one-electron or a part of a two-electron transition. Thus, a large number of terms in each sum in (2) vanishes without interaction that do not vanish with interaction. This can be visualized in figure 2 as an increase in the numerators therefore is to favor delocalization,.

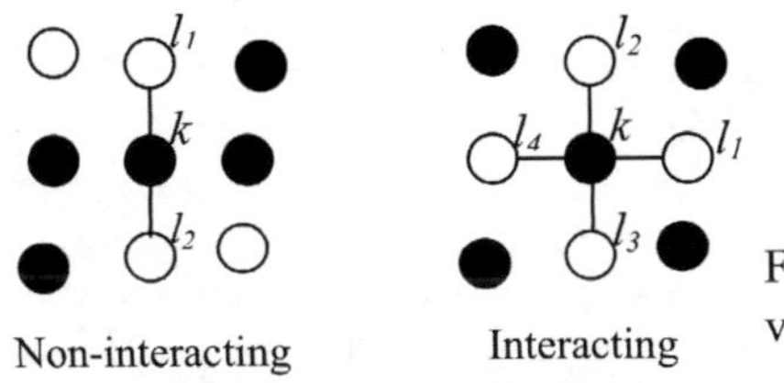

Figure 2. The effect of interaction on the number of non vanishing terms in $\sum_L\langle K_0|V|L_0\rangle$.

In summary, interaction increases both the numerators and the denominators. Bringing on divergence depends on whether there is a sufficient number of new paths introduced by interaction which do not decay rapidly with the perturbation order. The ratios $\sum_L\langle K_0|V|L_0\rangle/E_k\text{-}E_L$ can increase or decrease with interaction – an increase in their value enhances divergence, i.e. delocalization and a decrease enhances localization.

When the interactions are of Coulomb type the values of the transition matrix elements are

$$\langle I_0|V|J_0\rangle = r_{ij}^{-1}[1+W_{ij}^{-1}\textstyle\sum_k(r_{jk}^{-1} - r_{ik}^{-1})]\,\exp(-r_{ij}/\xi) \qquad (5)$$

when I_0 and J_0 differ in two sites and

$$\langle I_0|V|J_0\rangle = r_{ij}^{-1}\,r_{kl}^{-1}[1+W_{ij}^{-1}W_{kl}^{-1}(r_{jl}^{-1} + r_{ik}^{-1} - r_{jk}^{-1} - r_{il}^{-1})]\exp(-r_{ij}/\xi)\,\exp(-r_{kl}/\xi) \qquad (6)$$

when they differ in four sites. In (4) and (5) i is a site in I_0, j in J_0, k runs over all other sites identically occupied in I_0 and J_0 and $W_{ij} = \varepsilon_i\text{-}\varepsilon_j$. In (5) $k\in I_0$ and $l\in J_0$.

3. Quantitative consideration.

Two limits are considered here

- when the non-interacting system is *just* localized.
- when the non-interacting system is *strongly* localized

In the first case a necessary and sufficient condition for interaction to induce delocalization is

$$\sum_L\langle K|V|L\rangle/(E_I - E_L)|_{\text{int}} > \sum_L\langle K|V|L\rangle/(E_I - E_L)|_{non-\text{int}} \qquad (7)$$

i.e. for the delocalizing effect of the numerators to prevail over the localizing effect of the denominators.

The difference between the denominators in the two expression in (7) arises because of the interaction energy of a site with other sites in the system. Considering the interactions on a site as a random energy and the relation $W/E_C \approx 1$ associated with incipient localization, the denominators differ by a factor of 2.

The numerators are responsible primarily for a difference in the number of terms in the summation. For the condition of incipient localization there are about twice as many one-electron paths originating at a site in the interacting system as there are in the non-interacting system (see figure 2). This is true in 3D as well as in 2D. In addition, each such path can be traversed by a two-electron transition or a one-electron transition. So a good estimate is that there are four times the possible paths in the interacting system as in the non-interacting system, i.e. four times as many non-vanishing terms in the sum. This is actually a conservative estimate since there is a large number of two electron transitions containing each one-electron transition to l indicated in figure 2 (sites for the second electron transition are outside the figure). However, even with the conservative estimate (6) is clearly satisfied indicating that interactions *can* induce delocalization when the non-interacting system would still be localized.

In the case of strong localization $W >> E_C$. As a result the near neighbor correlation portrayed by figure 2 is very weak so interaction hardly changes the number of terms in the sum $\sum_L \langle K_0|V|L_0\rangle$. In addition, the effect of the interaction energies on the random energy is also very small, $W+E_C \approx W$. The total effect of interaction on localization in this regime is therefore very small.

4. Conclusions.

Interaction affects Anderson localization in two ways. One is to enhance localization the other to enhance delocalization. The first can be traced physically to an effective increase of the disorder energy, the latter to a correlation in the occupation and to a correlated motion of electrons. Near the Anderson transition the correlation effects prevail and interaction can cause delocalization of some states. The motion of the electrons is however highly collective so one can expect strong collective slowing of motion implying that conductivity at T=0 may be immeasurably small. If the energy regime of delocalized electron reaches the lowest energies, they can however, act as a thermal bath.

The reasoning leading to the conclusion that interactions can cause delocalization did not differentiate qualitatively between 2D and 3D. It is widely accepted that in the absence of interactions no truly delocalized states exist in the thermodynamic limit [9]. However the states can be localized weakly. One can thus conclude that in such cases interaction should bring aboutdelocalization.

Finally, it should be stressed that the considerations in this paper rest heavily on the Rayleigh Schrödinger perturbation expansion which is not an exact theory. The conclusions thus are not totally irrefutable.

Acknowledgements: The work profited greatly from discussions with the participants at the 2011 IAS workshop on electron glasses in Jerusalem and at the program Electron Glasses at KITP in Santa Barbara. Special thanks for stimulating discussions to Boris Shapiro and to Markus Müller and to my long time collaborator Zvi Ovadyahu for never tiring of arguing with me.

References

[1] Anderson P.W. 1958, *Phys. Rev.* **109** 1492
[2] Epperlein F., Schreiber M. and Vojta T. 1997, *Phys. Rev. B* **56** 5890
[3] Pollak M. and Knotek M.L. 1979 *J. Non-cryst. Solids* **32** 141, 1979
M. Pollak 1980 *Phil. Mag. B* **42** 781
[4] Basko D.L., Aleiner I.L. and Altshuler B.L. *Annals of Physics* **321** 11265201
[5] Fleishman L and Anderson P.W. 1980, *Phys. Rev. B* **21** 2366
[6] Müller M. and Ioffe L.2007, arXiv 0711.2668v1
[7] Pollak M. 1997 *Phys. Status Solidi* **205** 35
[8] Efros A.L. and Shklovskii B.I. 1975, *J. Phys. C* **8** L49
[9] Abrahams E. et. al., 1979, *Phys. Rev. Lett.* **42** 673

Journal of Physics: Conference Series **376** (2012) 012013 doi:10.1088/1742-6596/376/1/012013

Anderson localization in atomic mixtures

K. Ziegler[1], O. Fialko[2]

[1]Institut für Physik, Universität Augsburg, D-86135 Augsburg, Germany

[2]Centre for Theoretical Chemistry and Physics, Massey University (Albany Campus), Private Bag 102904, North Shore MSC, Auckland 0745, New Zealand

E-mail: klaus.ziegler@physik.uni-augsburg.de

Abstract. A mixture of two types of atoms in an optical lattice is studied. Assuming that one type of atoms is much heavier than the other, the complex interplay of the two species leads to a correlated distribution of heavy atoms and to diffusion or Anderson localization of the light atoms. The latter depends on the distribution of the heavy atoms, where diffusion appears only at low temperatures.

1. Introduction

In search for a physical system where we can manipulate disorder systematically, a gas of ultracold atoms turns out to be a good candidate. Ultracold atoms in an optical or magnetic trap have been used to simulate various many-body systems. The main advantage of the trapped ultracold gas is that most of the physical parameters can be tuned easily during the experiment. First of all, the quantum statistics, i.e. bosons or fermions, can be chosen by using the proper atomic species. For instance, ^{87}Rb atoms are bosons and ^{6}Li, ^{40}K atoms are fermions. The interaction between the atoms can be changed from attraction to repulsion by an external magnetic field due to a Feshbach resonance [1, 2]. The underlying lattice structure of a solid-state system can be provided in the atomic system by an optical lattice [4]. In such a structure the atoms tunnel from one potential well to a neighboring one just like the electrons in a metal. On the other hand, the optical potential, which is created by counter-propagating laser fields, can easily be modified in strength by changing the laser intensity. Consequently, the tunneling rate of the atoms can be tuned by the latter. This has been used to study the Bose-Einstein condensate to Mott state transition in a Bose gas, where a Bose-Einstein condensate appears at large tunneling rates and a Mott phase with commensurate filling of the optical lattice at small tunneling rates [4].

Disorder in statistical or solid-state physics usually means that particles are scattered by a disordered (i.e. a randomly produced) array of other particles. This disordered array is static and does not change during the course of an experiment. For instance, the scattering of electrons in a crystal with randomly distributed impurity atoms is a typical realization of disordered physics. A comparable situation can be constructed in atomic systems subject to an optical lattice. The periodic structure of the optical lattice is normally quite robust such that disorder in the sense of impurities cannot easily be produced. However, in contrast to a crystal in solid-state physics, two or more different types of atoms can be mixed and brought in an optical lattice, where one type of the atoms is relatively light (e.g. ^{6}Li) and the other type(s) is (are) heavy (e.g. ^{40}K, ^{23}Na [5], ^{87}Rb). The different masses lead to different dynamical properties. In particular, the

Published under licence by IOP Publishing Ltd

heavy atoms behave almost like static particles, whereas the light atoms tunnel freely in the optical lattice. Consequently, the heavy atoms play the role of the impurity atoms in the crystal and the light atoms the role of the electrons. Thus disorder physics can be studied in mixtures of atoms which are available in experiments [6].

Disordered systems have attracted a lot of attention because they provide a new class of physics, including new phases and new types of phase transitions. The main reason is that scattering of quantum particles by a periodic structure is qualitatively different from scattering by a disordered structure: while the former is governed by the Bloch theorem, the latter experiences diffusion (for weak disorder) or even Anderson localization for strong disorder [7]. There is a phase transition between the diffusion-controlled regime and the localized regime [8, 9].

A fundamental concept for a theoretical description of disordered system is that physical quantities are averaged with respect to a statistical distribution of the random structures. It will be discussed in this article that averaging over thermal fluctuations of heavy atoms provides such a disorder (or quenched) average of physical quantities, like the density or the mean-square displacement of light atoms.

2. Model

The prototype of fermionic many-body lattice systems is the Hubbard model which describes tunneling of spin-1/2 fermions (e.g. electrons) between nearest-neighbor lattice sites and a local interaction. This model can be generalized to describe the interaction of spinless fermion atoms with localized atoms as

$$H = -t_f \sum_{\langle r,r' \rangle} f_r^\dagger f_{r'} + \sum_r [-\mu_b b_r^\dagger b_r - \mu_f f_r^\dagger f_r + U_{bf} b_r^\dagger b_r f_r^\dagger f_r + U_{bb} b_r^\dagger b_r^\dagger b_r b_r] \ . \tag{1}$$

Here $f^\dagger(f)$ are creation (annihilation) operators of the light fermionic atoms, $b^\dagger(b)$ are the corresponding operators of the heavy atoms (fermions or bosons). The interaction between the light fermions at low temperatures is controlled by the Pauli principle, whereas the interaction between light fermions and the heavy atoms is controlled by U_{bf} and the interaction between the heavy atoms by U_{bb}. The chemical potentials μ_b and μ_f controls the number of heavy and light atoms, respectively.

A grand-canonical ensemble is defined by the partition function, which is the trace of Boltzmann weights over all realizations of heavy and light atoms in the system:

$$Z = Tr_b Tr_f e^{-\beta H} \ . \tag{2}$$

Since there is no tunneling for the heavy atoms, they are characterized only by integer occupation numbers: for bosons $n_r = 0, 1, 2, 3, ...$ and for fermions due to the Pauli principle $n_r = 0, 1$. After performing the trace over the light fermions we arrive at the partition function

$$Z = \sum_{\{n_r\}} e^{\beta \sum_r [\mu_b n_r - U_{bb} n_r (n_r - 1)]} \det \left[\mathbf{1} + e^{-\beta h_f} \right] \ , \tag{3}$$

where h_f is a matrix with respect to the lattice sites that describes tunneling of light atoms and includes a one-particle potential. The matrix elements are $(h_f)_{r,r'} = -\hat{t}_{r,r'} + (U_{bf} n_r - \mu_f)\delta_{r,r'}$.

2.1. Characterization of heavy atoms

According to Eq. (3) the heavy particles are distributed according to the probability distribution

$$P(\{n_r\}) = \frac{1}{Z} e^{\beta \sum_r [\mu_b n_r - U_{bb} n_r (n_r - 1)]} \det \left[\mathbf{1} + e^{-\beta h_f} \right] , \tag{4}$$

with $\sum_{\{n_r\}} P(\{n_r\}) = 1$. Thus the density of heavy atoms at lattice site r is given by $\langle n_r \rangle_b = \sum_{\{n_r\}} n_r P(\{n_r\})$.

2.2. Characterization of light atoms

The characterization of the light atoms is controlled by quantum effects (i.e. tunneling between the potential wells of the optical lattice) and the interaction with the classical heavy atoms. Thus we need to describe the dynamics of the light atoms by the evolution of their quantum state.

In general terms, we consider a quantum gas in the initial equilibrium state $|\Psi_0\rangle$ and create a new state by adding a light atom at site 0: $c_0^\dagger|\Psi_0\rangle$. Then we follow the evolution of the state with the additional particle, controlled by the Hamiltonian H, until the time t, where we get the (non-equilibrium) state $|\Phi_t\rangle = e^{-iHt}c_0^\dagger|\Psi_0\rangle$. From this we obtain for the mean-square displacement

$$\sum_r r^2\langle\Phi_t|c_r^\dagger c_r|\Phi_t\rangle = \sum_r r^2\langle\Psi_0|c_0 e^{iHt}c_r^\dagger c_r e^{-iHt}c_0^\dagger|\Psi_0\rangle \ . \tag{5}$$

Instead of using a fixed state $|\Psi_0\rangle$ we average with respect to a Boltzmann ensemble at inverse temperature β of energy eigenstates:

$$\langle r^2\rangle = \sum_r r^2 \frac{\sum_k e^{-\beta E_k}\langle E_k|c_0 e^{iHt}c_r^\dagger c_r e^{-iHt}c_0^\dagger|E_k\rangle}{\sum_k e^{-\beta E_k}} = \frac{1}{Z}\sum_r r^2 \mathrm{Tr}\left[e^{-\beta H}c_0 e^{iHt}c_r^\dagger c_r e^{-iHt}c_0^\dagger\right] \ . \tag{6}$$

where the right-hand side can also be expressed by the density of light atoms N_r as $\sum_r r^2 N_r$. For long times and weak coupling U_{bf} (or low temperatures) diffusion is expected with $\langle r^2\rangle \sim Dt$ and the diffusion coefficient D. On the other hand, for stronger coupling (or higher temperatures) Anderson localization is expected. Thus there is no expansion at all, and the mean-square displacement is finite for $t \sim \infty$. Intermediate regimes with sub-diffusive behavior are also possible, where $\langle r^2\rangle \sim Dt^\alpha$ with $0 < \alpha < 1$.

Returning to our specific case of the atomic mixture, we can separate the heavy atoms and the light atoms as in the partition function (3). This allows us to write for the mean-square displacement [10]

$$\langle r^2\rangle = \frac{1}{Z}\sum_{\{n_r\}} e^{\beta\sum_r[\mu_b n_r - U_{bb}n_r(n_r-1)]}\det\left[\mathbf{1}+e^{-\beta h_f}\right]$$
$$\times\sum_r r^2[e^{ith_f}(\mathbf{1}+e^{-\beta h_f})^{-1}]_{0r}[e^{-ith_f}(\mathbf{1}+e^{-\beta h_f})^{-1}]_{r0}. \tag{7}$$

In the density $\langle n_r\rangle_b$ and in the mean-square displacement there is the distribution density of Eq. (4) as a common non-negative factor. This implies that density of light atoms N_r and their mean-square displacement $\langle r^2\rangle = \sum_r r^2 N_r$ appear as expectation values with respect to the distribution density of Eq. (4) as

$$N_r = \langle [e^{ith_f}(\mathbf{1}+e^{-\beta h_f})^{-1}]_{0r}[e^{-ith_f}(\mathbf{1}+e^{-\beta h_f})^{-1}]_{r0}\rangle_b \ . \tag{8}$$

Thus the distribution of heavy atoms is a realization of quenched correlated disorder.

Using the single-particle Green's function

$$\mathcal{G}_{rr'}(t) = [e^{-ith_f}(\mathbf{1}+e^{-\beta h_f})^{-1}]_{rr'} \ , \tag{9}$$

we obtain $N_r = \langle\mathcal{G}_{0r}^\dagger(t)\mathcal{G}_{r0}(t)\rangle_b$. For a given configuration $\{n_r\}$ of heavy atoms the Green's function can also be expressed by eigenfunctions of the single-particle Hamiltonian h_f ($h_f\phi_k = e_k\phi_k$). The spatial properties of these eigenfunctions determine the spreading of the average particle density N_r through the Green's function:

$$\mathcal{G}_{r0}(t) = \sum_k e^{-ie_k t}\frac{\phi_{k,r}^*\phi_{k,0}}{1+e^{-\beta e_k}} \ . \tag{10}$$

The denominator represents the Fermi-Dirac function, reflecting the fact that our atoms are fermions. At low temperatures all states with $e_k > 0$ (i.e. states with energy below the chemical potential according to the definition of h_f) contribute equally to the Green's function. According to the theory of Anderson localization, it can be assumed that $|\phi_{k,r}| \sim e^{-|r|/\xi_k}$, where ξ_k is the localization length. After a Fourier transformation of the time-dependent density in Eq. (8), the $\omega = 0$ Fourier component of N_r reads

$$\int_0^\infty N_r dt = \sum_k \frac{|\phi^*_{k,r}\phi_{k,0}|^2}{(1+e^{-\beta e_k})^2} \sim \frac{e^{-2|r|/\xi}}{(1+e^{-\beta e_{k_0}})^2} \quad (r \sim \infty) , \tag{11}$$

where ξ is the largest localization length and e_{k_0} the corresponding energy level. Thus the expansion of the wave packet on large scales is controlled by ξ. This result suggests that the spatial expansion of an atomic cloud is governed by the largest length scale of the system.

3. Strong-coupling expansion: classical Ising model

In order to evaluate the distribution of heavy atoms and the localization length of the light atoms, we employ an approximation to the distribution function. The simplest case is a weak-coupling expansion for $U_{bf}/t_f \ll 1$. However, this leads only to a renormalized chemical potential [11] and is not very useful here. A strong-coupling expansion with $t_f/U_{bf} \ll 1$, on the other hand, provides an effective coupling of heavy atoms between nearest-neighbor potential wells [11, 12, 13, 14]. In leading order of this expansion the determinant in Eq. (3) reduces to (cf. [15])

$$\det\left[\mathbf{1} + e^{-\beta h_f}\right] \approx e^{\frac{\beta}{2}\sum_{\langle r,r'\rangle} E(n_r,n_{r'})} \prod_r \left[1 + e^{\beta A_r}\right] \tag{12}$$

with $A_r = \mu_f - U_{bf} n_r$ and

$$E(n_r, n_{r'}) = \frac{e^{\beta A_r} - e^{\beta A_{r'}}}{A_r - A_{r'}} \frac{t_f^2}{(1+e^{\beta A_r})(1+e^{\beta A_{r'}})} . \tag{13}$$

Then the partition function in Eq. (3) reads

$$Z \approx \sum_{\{n_r\}} e^{-\beta \sum_r F(n_r)} e^{\frac{\beta}{2}\sum_{\langle r,r'\rangle} E(n_r,n_{r'})} , \tag{14}$$

where

$$F(n) = -\frac{1}{\beta}\log\left[1 + e^{\beta(\mu_f - U_{bf} n)}\right] - (\mu_b + U_{bb})n + U_{bb} n^2 . \tag{15}$$

As the temperature is reduced, the bosons organize themselves in a low-energy state. Beginning with a small tunneling rate t_f we can neglect the second factor in the Boltzmann weights of Eq. (14) and determine the number $\bar{n}$ which minimizes $F(n)$. At least one minimum exists because $F(n)$ increases monotonously for large values of n. Assuming that the value for the minimum of $F(n)$ is a number between the two integers $[\bar{n}]$ and $[\bar{n}] + 1$, we study small fluctuations around $\bar{n}$ with $n_r = [\bar{n}]$ and $n_r = [\bar{n}] + 1$. This enables us to rewrite the function $E(n_r, n_{r'})$, since only four different values are possible for $S_r = 2(n_r - [\bar{n}]) - 1$ with $S_r = \pm 1$:

$$E(n_r, n_{r'}) = E_0 + E_1(S_r + S_{r'}) + E_2 S_r S_{r'} , \tag{16}$$

where the coefficients, using the short-hand notation $\epsilon(S, S') = E([\bar{n}] + (S+1)/2, [\bar{n}] + (S'+1)/2)/4$, are

$$E_0 = \epsilon(1,1) + \epsilon(-1,-1) + 2\epsilon(1,-1), \quad E_1 = \epsilon(1,1) - \epsilon(-1,-1), \quad E_2 = \epsilon(1,1) + \epsilon(-1,-1) - 2\epsilon(1,-1) . \tag{17}$$

Journal of Physics: Conference Series **376** (2012) 012013 doi:10.1088/1742-6596/376/1/012013

Figure 1. The distribution of heavy bosons at half-filling for $U_{bb} = U_{bf} = 10t_f$. The left plot is for $k_BT = 0$, where bosons form perfect chess-board-like pattern. The middle plot is for $k_BT = 2t_f$, where the heavy bosons form chess-board-like clusters and clusters with more homogeneous patterns. The right plot is for $k_BT = 8t_f$: at some lattice sites there are more than one boson as compared to the previous cases.

Eq. (16) provides an interpretation in terms of a classical Ising model for the spin $S_r = \pm 1$: E_1 is proportional to an external magnetic field and E_2 is the nearest-neighbor spin-spin coupling. Closer inspection reveals that the latter is negative, which indicates an antiferromagnetic coupling. This produces a staggered configuration with alternating $n_r = [\bar{n}], [\bar{n}]+1$ occupation for small E_1. Then there is also competition between the antiferromagnetic spin-spin coupling and the magnetic field term $E_1 S_r$, where the latter favors ferromagnetic order with homogeneous occupation. This implies a first-order transition between homogeneous and staggered (checkerboard in $d = 2$) order of the heavy atoms. In the proximity of this first-order transitions there is phase separation, leading to the coexistence of homogeneous and staggered clusters. This is shown in the second plot of Fig. 1.

4. Results and conclusions

distribution of heavy atoms: The strong-coupling expansion, leading to an Ising model with antiferromagnetic coupling and an effective magnetic field, indicates already a complex distribution of heavy atoms. Now we employ a numerical approach to study the distribution in more detail and to evaluate the finite-size properties of localization length ξ of the light fermions in Eq. (11). For this purpose we consider a two-dimensional lattice with $\mathcal{N} = N \times N$ ($N = 20$) lattice sites and derive the statistically relevant realizations by Monte-Carlo sampling of the distribution of $\{n_r\}$. Examples at different temperatures are presented in Fig. 1.
expanding cloud of light atoms: An initial state of light atoms with high density at the center is prepared. Evolution under the effect of e^{-iHt}, as described in Sect. 2.2, leads either to diffusive-like expansion (cf. left panel in Fig. 2) or stays approximately in the initial state (cf. right panel in Fig. 2), depending on the strength of disorder created by the heavy atoms.
Localization transition of light atoms: The localization length ξ is studied under the change of length scales of a finite lattice of length L and width M [9], representing an adiabatically expanding atomic cloud. In particular, we analyze the change of the localization length with respect to the width M. For this purpose, we define a reduced localization length of light atoms as $\Lambda_M = \xi/M$ and calculate this quantity by means of a numerical transfer-matrix approach for a given realization $\{n_r\}$ of heavy atoms [16]. For the latter statistically relevant realizations are chosen by Monte-Carlo sampling of the distribution of heavy atoms. Since for sufficiently large values of L (we use typically $L \sim 10^8$) the localization length is presumably self-averaging, there is no need for additional averaging over different realizations. We use open boundaries along the action of the transfer matrix and periodic boundary conditions in the perpendicular direction.

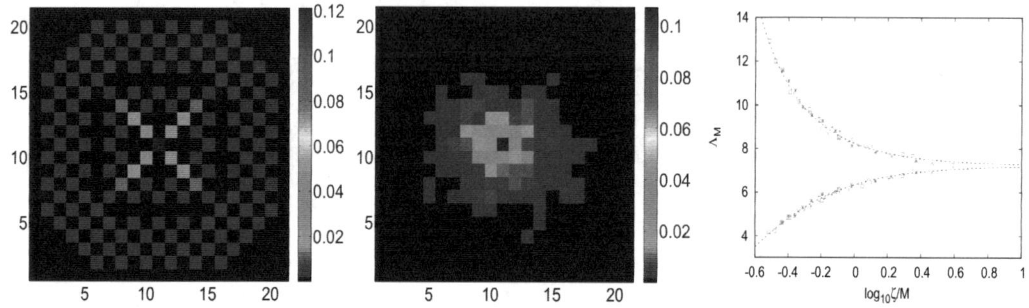

Figure 2. Evolution of a cloud of light atoms after $t = 10\hbar/t_f$. Left panel: expansion of the wave packet for weak disorder at $k_BT = 0.05t_f$. Middle panel: localization of the wave packet for strong disorder at $k_BT = 0.2t_f$. Right panel: Scaling of the localization length near the Anderson transition (from Ref. [18]).

Λ_M either increases (delocalized states) or decreases (localized states) with the width M, depending on the system parameters (e.g. the inverse temperature β). There can also be a marginal behavior (e.g. for a special value β_c'), where Λ_M does not change with M. The latter indicates the existence of a phase transition from localized to delocalized states. A quantitative description of the behavior near β_c' can be based on the one parameter scaling hypothesis [9, 17]: Using the parameter $\zeta = |\beta - \beta_c'|^{-\nu}$, we can fit our numerical data for Λ_M quite well with

$$\Lambda_M = \Lambda_c \exp\left[\pm A\left(\frac{\zeta}{M}\right)^{-1/\nu}\right] , \tag{18}$$

with $\Lambda_c \approx 10.9$, $A \approx 0.09$, and $\nu \approx 0.88$, as shown in the right panel of Fig. 2 [18].

From these results we can conclude that (i) a mixture of two atomic species creates a correlated distribution of heavy atoms and (ii) the light atoms undergo an Anderson transition, where localized (delocalized) states appear at high (low) temperatures.

[1] Fano U 1935 *Nuovo Cimento* **156** 12
[2] Feshbach H 1958 *Ann. Phys. (N.Y.)* **5** 357
[3] Kleppner D 2004 *Physics Today* **57** 12
[4] Greiner M, Mandel O, Esslinger T, Hänsch Th W and Bloch I 2002 *Nature* **415** 39
[5] Stan CA, Zwierlein MW, Schunck CH, Raupach SMF and Ketterle W 2004 *Phys. Rev. Lett.* **93** 143001
[6] DeMarco B and Jin DS 1999 *Science* **285** 1703
[7] Anderson PW 1958 *Phys. Rev.* **109** 1492
[8] Wegner F 1976 *Z. Phys. B* **25** 327
[9] Abrahams E, Anderson PW, Licciardello DC and Ramakrishnan TV 1979 *Phys. Rev. Lett.* **42** 673
[10] Ziegler K 2006, *Laser Physics* **16** 699
[11] Ates C and Ziegler K 2005 *Phys. Rev. A* **71** 063610
[12] Falicov LM and Kimball JC 1969 *Phys. Rev. Lett.* **22** 997
[13] Farkasovsky P 1997 *Z. Phys. B* **102** 91; 2008 *Europhysics Lett.* **84** 37010
[14] Freericks JK and Zlatić V 2003 *Rev. Mod. Phys.* **75** 1333
[15] Ziegler K and Schneider D 2005 *J. Phys.: Condens. Matter* **17** 5489
[16] Pichard JL and Sarma G 1981 *J. Phys. C* **14** L127; Slevin K and Ohtsuki T 1999 *Phys. Rev. Lett.* **82** 382
[17] MacKinnon A and Kramer B 1983 *Z. Phys. B* **53** 1
[18] Fialko O and Ziegler K 2009 *Europhys. Lett.* **85**, 60003

The destruction of the virtual Anderson transition in a narrow impurity band of doped quantum well structures.

N. V. Agrinskaya, V. I. Kozub, D. S. Poloskin
Ioffe Physical-Technical Institute of the Russian Academy of Sciences, 194021, Saint Petersburg, Russia.
E-mail: nina.agrins@mail.ioffe.ru

Abstract.
Earlier we reported an observation at low temperatures of activation conductivity with small activation energies in strongly doped uncompensated layers of p-GaAs/AlGaAs quantum wells. We attributed it to Anderson delocalization of electronic states in the vicinity of the maximum of the narrow impurity band due to the small width of the impurity band characterized by weak disorder. In this case the carriers were activated from the "band tail" while its presence was related to weak background compensation. Here we study an effect of the extrinsic compensation and of the impurity concentration on this "virtual" Anderson transition. It was shown that an increase of compensation initially does not affect the Anderson transition, however at strong compensations the transition is suppressed due to increase of disorder. However at larger impurity concentration the conductivity becomes to be metallic due to Mott transition.

1. Introduction

In our previous papers [1], [2] we reported an observation of delocalized electronic states in the vicinity of the maximum of a narrow impurity band in GaAs/AlGaAs quantum well structures. It is important that such a delocalization took place at dopant concentrations significantly smaller than ones corresponding to the Mott-Anderson criterion. Such a behavior we described as a manifestation of the virtual Anderson transition - because, despite of a presence of delocalized states, transport of the majority carriers over these (occupied) states was blocked by the Hubbard correlations. The smallness of the critical concentration for the Anderson transition was explained by a weakness of the random potential and, correspondingly, by a smallness of the scatter of localized states energies. In this case linear carrier transport was supported by minority carriers activated from acceptors ionized by background impurities. Note that such a behavior can be realized only 2D systems, where the background compensating defects can be situated outside of the 2D layer which is important for a formation of the narrow impurity band. Since the conductivity was related to a presence of finite (but weak) compensation and to activation of the minority carriers situated in the band tail to the band of delocalized states, the corresponding activation energy was by an order of magnitude smaller than the activation energy of the dopant. Another mechanism of conductivity over the delocalized states was relevant for the case of strong enough electric fields. It was related to an impact ionization of the minority carriers to the band of delocalized states. According to the considerations given

Journal of Physics: Conference Series **376** (2012) 012014 doi:10.1088/1742-6596/376/1/012014

above, our picture corresponds to the case when the Fermi level is situated below the mobility edge within the impurity band. One could expect that an increase of a compensation degree, which shifts the Fermi level towards the mobility edge, finally would lead to purely metallic (non-activated) conductivity. However,significant increase of the compensation can lead to an increase of disorder, i.e. to an increase of the critical concentration, and, finally, to a suppression of the delocalization.

In its turn, an increase of this concentration can lead to an overlap of the two Hubbard bands and to appearance of the charged centers not related to compensation. Again, the additional charge disorder can suppress the delocalization. However the further increase of the dopant concentration finally leads to the Mott transition when the Fermi level reaches the mobility edge of the upper Hubbard band. Thus, it was of interest to study an effect of both compensating centers concentration (introduced artificially) and of the dopant concentration on the manifestation of the virtual Anderson transition. It is this investigation which is reported in this paper. We will show that an increase of compensation initially leads to a decrease of the activation energy in agreement to the considerations given above. The further increase of compensation leads to a complete suppression of the virtual Anderson transition which is accompanied by a significant broadening of the impurity band. The latter manifests itself in an increase of the all activation energies. Then, an increase of the dopant concentration in uncompensated samples also initially leads to a suppression of the virtual Anderson transition. However at large enough concentrations the metallic state resulting from the Mott transition is formed.

2. Experiment

We have studied two sets of $GaAs/Al_{0.3}Ga_{0.7}As$ quantum well samples, grown by molecular epitaxy technique. The samples contained 5 quantum wells with a width 15 nm separated by $Al_{0.3}Ga_{0.7}As$ barriers with a thickness 100 nm. At the first set of samples only quantum wells centers (5 nm) were doped by Be with large enough concentration $(1-3)\ 10^{12}\ ^{-2}$ (samples N 1,2). At the second set of samples the centers of the quantum wells were also doped by Be, however the centers of barriers (5 nm) were doped by compensating impurity (Si), see Table 1 (samples 3,4). Thus, the degrees of compensation $K = N_D/N_A < 0,01$ were less than 0.01 for samples 1,2 and for samples 3 and 4 K=0.1 and K=0.5, respectively. The acceptor concentration in the wells was taken to be large enough to ensure a presence of delocalized states in the center of impurity band of A_0 centers (resulting from the virtual Anderson transition)[1], [2].

N	N_A, cm^-2	N_D, cm^-2	p_{300K}, cm^{-2}	ε_1 meV	ε_4,meV	$K = N_D/N_A$
1-7-582	$2,5 \cdot 10^{12}$	-	$2 \cdot 10^{12}$	—	-	≤ 0.01
2-7-580	$1,5 \cdot 10^{12}$	,	$1 \cdot 10^{12}$	21	2	≤ 0.01
3-8-291	$1 \cdot 10^{12}$	Si, $1,5 \cdot 10^{11}$	$1 \cdot 10^{12}$	21	1,5	0.1
4-8-292	$1 \cdot 10^{12}$	Si, $5 \cdot 10^{11}$	$5 \cdot 10^{11}$	40	10	0.5

Table 1. Parameters of the studied samples

On Fig.1(a,b) we give the temperature curves of the conductivity σ and of carrier concentration P (obtained from the Hall coefficient R_h) for all of the samples.

When only centers of the wells are doped, there exists the filled impurity band of the singly occupied acceptors A_0, where the compensation degree K is small enough and is controlled by a presence of random donor impurities in the barrier ($N_D < 10^{16}$ cm^{-3}). The delocalized states appear in this band at concentration $N_A \sim 10^{18}\ cm^{-3}$.In this case the concentration of carriers within the allowed band a standard expression for weakly compensated impurity can be used. Such a behavior is observed for doped quantum wells without intended compensation or at weak compensation (samples 2 and 3).

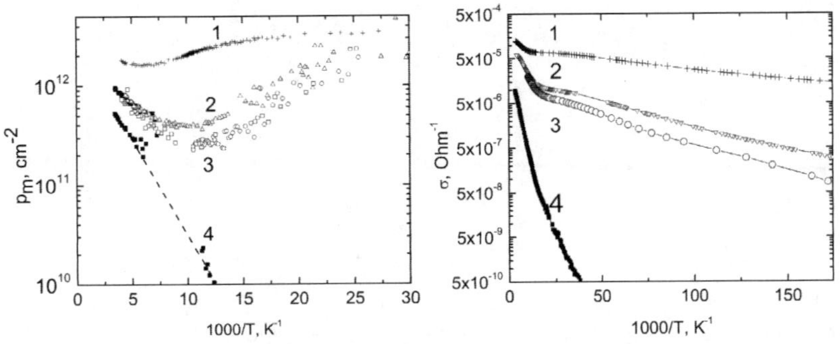

Figure 1. Temperature dependence of the (a) conductivity and (b) hole concentration for all samples listed in the table.

As it was noted in our previous paper [3], one of the features characterizing the virtual Anderson transition in the impurity band is a mixed conductivity (over the allowed and impurity bands) which manifests itself as a presence of a characteristic minimum in temperature behavior of concentration. At higher temperatures there exists a part of the curve corresponding to activation of the carriers from the Fermi level to the allowed band - energy ε_1. This minimum appears to be deeper with a sharp increase of the effective concentration at temperatures 20-50 K provided the carriers in the allowed band and in the impurity band have opposite charges. When calculating the mixed concentration P_m from the Hall effect one should take into account that in the impurity band completely filled by holes the carriers can be of the opposite sign [3].

Note that P_m can diverge (tend to infinity) when the concentrations of electrons in the impurity band and concentration of holes in the valence band are nearly equal. Thus the Hall concentration at some temperatures can appear to be larger, than the concentration at room temperature. The mixed conductivity (in contrast to mixed concentration) has no maxima (at least if one does not take into account temperature behavior of the mobility) and manifests itself as a shoulder in the conductivity temperature curves. At the lower temperatures the temperature behavior is controlled by the activation energy ε_4 (related to an activation of the carriers from the Fermi level to the mobility edge within the impurity band). At the weak compensations this energy is about a half width of the impurity band. At temperatures less than 4 K, these samples demonstrated the breakdown behavior related to the impact ionization of the carriers from the Fermi level to the mobility edge.

An important difference between the weakly compensated sample N3 from the uncompensated N2 is less value of the activation energy ε_4, as well as the breakdown behavior at higher temperatures (around 4 K), see Fig.2 (note that the measurements were made at constant current 1 nA).

In the sample with a strong compensation degree N4 ($N_D/N_A \sim 0,5$) the mixed conductivity as well as activated temperature behavior at small temperatures and the breakdown behavior were not observed. It is related to the significant increase of disorder and, correspondingly, to the increase of the width of the impurity band which leads to a suppression of the Anderson transition. The activation energy ε_1 (obtained from the Hall data) in this sample is $\sim$ 40 meV which exceeds significantly the corresponding energies for weakly compensated samples. In its turn, the low temperature conductivity behavior demonstrates activation energy $\varepsilon_3 \sim$ 10 meV.

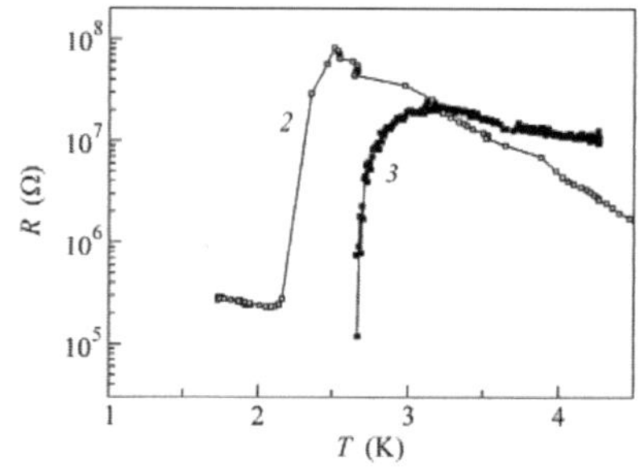

Figure 2. Temperature dependence of the low-temperature conductivity of samples 2 and 3 measured in the constant current regime at the current of 1 na.

If one relates this energy to the nearest neighbor hopping, it corresponds to the half width of the impurity band. An increase of the energy ε_1 for the strong compensation limit, according to [4], is equal to $\varepsilon_1 = \varepsilon_0 + \varepsilon_3$, which for $\varepsilon_0 = 30$ meV gives the observed energy $\varepsilon_1 = 40$ meV. The variable range hopping was not observed due to very large values of resistance at low temperatures.

With an increase of the doping level the sample N1 demonstrates weaker temperature dependencies of the conductivity and Hall effect (Fig.1). It is important that the sign of the Hall effect at room temperature and at helium temperature is the same, that is the type of carriers does not change with temperature decrease. In its turn, the Hall effect gives the same concentration of holes at these temperatures. Such a behavior contradicts with the one corresponding to the virtual Anderson transition and evidences the standard Mott transition. This sample exhibits effects of weak localization in temperature behavior of conductivity as well as in magnetoresistance (MR) (Fig.3). At low temperatures in weak magnetic fields MR appears to be positive (antilocalization), then a transition to negative MR is observed (weak localization). This behavior correlates to the one of quasimetallic samples (see [5]). Such a behavior evidences the Mott transition (overlapping of the upper and lower Hubbard bands).

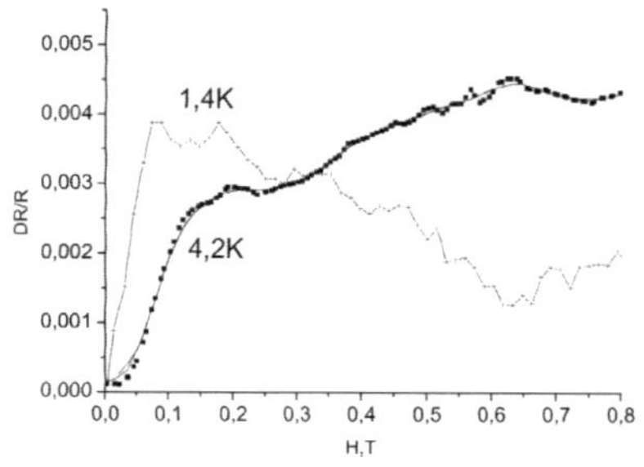

Figure 3. Magnetoresistance curves for sample 1

3. Discussion

1. Role of compensation.

As it was noted earlier, one could expect that an increase of compensation degree, shifting the Fermi level μ within the impurity band towards the band of delocalized states could

Journal of Physics: Conference Series **376** (2012) 012014 doi:10.1088/1742-6596/376/1/012014

finally to nullify the activation energy ε_4 which would mean a transition to metallic state. Indeed, assuming that the density of states of the impurity band has Gaussian shape, for small concentrations of the compensating donors one obtains

$$\int_{-\infty}^{\mu} g_0 \exp -(\frac{\varepsilon_0 - \varepsilon'}{\Delta\varepsilon_0})^2 d\varepsilon' = N_d \tag{1}$$

where ε_0 corresponds to a center of the impurity band, $\Delta\varepsilon_0$ is the halfwidth of the impurity band at the limit of small N_d. The latter is controlled mostly by deformational effects of non-Coulombic nature and by the order of magnitude coincides with energy ε_4 observed for weakly compensated samples. In thius case the shift of the Fermi level is given as

$$\frac{\partial\mu}{\partial n_d} = g_0^{-1} \exp(\frac{\varepsilon_0 - \mu}{\Delta\varepsilon_0})^2 \tag{2}$$

It is seen that for small N_d when the Fermi level is deep within the tail of the impurity band, an increase of N_d initially leads to strong shift of μ and, correspondingly, to a decrease of ε_4. However an increase of the Coulomb disorder leads to the broadening of the impurity band which can be estimated as

$$\Delta\varepsilon_C \sim \frac{e^2 N_d^{1/2}}{\kappa} \tag{3}$$

Correspondingly, finally this broadening can exceed the value of $\Delta\varepsilon_0$. According to considerations, given in our papers [1], [2], the criterion for Anderson transition for significantly compensated samples obtains a form

$$N_A a^2 \geq \frac{\alpha}{\ln(\varepsilon_0/(\Delta\varepsilon_C))} \tag{4}$$

where N_A is a concentration of dopant acceptors, a is the localization length while α is of the order of unity.

This equation can be compared with a criterion of Anderson transition in non-compensated samples :

$$N_A a^2 \geq \frac{\alpha}{\ln(\varepsilon_0/(\Delta\varepsilon_0))} \tag{5}$$

Since the mobility edge is expected to be situated in the vicinity of the band center, an increase of the compensation can shift the sample to the metallic state only at small degrees of compensation. When the Fermi level reaches a vicinity of the band center it inevitably means an icrease of $\Delta\varepsilon_C$ up to the values significantly exceeding ε_0. As it is seen from the comparison of the criteria 4and 5, this leads (for fixed concentration N_A) to a suppression of the virtual Anderson transition.

As it is seen from the estimate 3, at compensating donor concentrations $\sim 10^{-11}$ cm^{-2} the magnitude of the disorder potential is of the order of $\sim 6meV$. Thus in the sample 3 with a small degree of compensation with the concentration of the compensating donors given above the energy $\Delta\varepsilon_C$ appears to be comparable to the energy $\Delta\varepsilon_0$ (characterizing the bandwidth without compensation). Indeed, according to our estimates ([1], [2]) the energy $\Delta\varepsilon_0$ is of the order of $6meV$. This estimates is compatible to the activation energy $\varepsilon_4 \sim 2meV$.

This scenario completely agrees with experiment. In particular, for moderate degree of compensation ($K = 0.1$, sample 3) the criterion of the Anderson transition 4 is still met, and the observed picture is similar to the one of the sample 1. Note that despite of the expected shift of the chemical potential towards the center of the band, it does not lead to a significant decrease of the activation energy. This is due to the fact that the effect of the shift of μ

is compensated by the effect of the broadening of the very impurity band, as well as by a narrowing of the band of Anderson-delocalized states due to an increase of disorder. However one can not exclude a possibility that at some relation between the dopant concentration and the degree of compensation the real Anderson transition can occur in the impurity band. Namely, the real quasimetallic conductivity can be achieved when the Fermi level reaches the mobility edge. Indeed, one of our samples demonstrated such a behavior - at the dopant concentration $5 \cdot 10^{11}$ cm^{-2} and weak compensation the conductivity and magnetoresistance demonstrated weak localization behavior [5].

At the same time for strongly compensated sample 4 the activation energy ε_1 (obtained from the Hall measurements) is $\sim$ 40 meV which significantly exceeds the corresponding energies for weakly compensated samples. Then, at low temperatures the temperature behavior of conductivity exhibits activation energy $\varepsilon_3 \sim 10$ meV. If one attributes this energy to the nearest neighbor hopping, it also coincides with a halwidth of the impurity band.

Thus an absence of the virtual Anderson transition as well as a significant increase of the energies ε_1 and ε_3 at strong compensation is explained by a strong increase of the bandwidth due to the Coulomb effects. In this case $\varepsilon_1 = \varepsilon_0 + \varepsilon_3$, which for ε_0 =30 meV gives the observed energy ε_1=40 meV.

2. Role of the dopant concentration.

It is of some paradox that an increase of the dopant concentration (for a given small degree of compensation) should initially lead to a suppression of the virtual Anderson transition. Indeed, the increase of concentration leads to a broadening of the impurity band due to overlapping of the wavefunctions of the neighboring cites which we denote as ε_T. If $\Delta\varepsilon_T > \Delta\varepsilon_0$ one expects an overlap of the tails of the two Hubbard bands. This fact leads to an appearance of the charged states within the impurity band and, correspondingly, of the related disorder potential. The latter can be estimated by a standard expression 3, however instead of the concentration of the charged centers one should insert the concentration of doubly occupied N_A^+ centers resulting from the overlap of the Hubbard bands. Thus, according to 4 it initially leads to a suppression of delocalization. However one has in mind that the overlap of the Hubbard bands exponentially depends on the dopant concentration while the overlap mentioned above finally leads to the Mott transition and, correspondingly, to non-activated metallic conductivity. It is such a behavior which is observed for strongly doped sample 1. This sample demonstrates weak temperature behavior of conductivity and Hall effect (the sign of the Hall effect at room temperature and at helium temperature is the same, that is the type of carriers does not change with temperature decrease), while low-temperature MR demonstrates effects of weak localization and antilocalization at weak magnetic fields, Fig.3. This correlates with a behavior of dirty metals. As it was noted above. this behavior results from the Mott transition (that is from the strong overlap of the upper and lower Hubbard bands).

Thus we have shown that an increase of the compensation degree in the samples, demonstrating virtual Anderson transition, initially leads to an increase of the low temperature conductivity. However a further increase of the degree of compensation leads to a suppression of the transition (and, correspondingly, to a decrease of the conductivity) due to an increase of the disorder potential. In its turn, at large dopant concentration the Mott transition takes place, and the sample becomes to be metallic.

This work was supported by Russian Foundation for Basic Research (project no. 10-02-00544)and President grant NSH-3306.2010.2

[1] Agrinskaya N.V., Kozub,V.I., Poloskin D.S. Jetp Lett., **85**, 169 (2007)
[2] N. V. Agrinskaya, V. I. Kozub, Yu. M. Galperin, D. V. Shamshur J. Phys.C 20, 395216 (2008)
[3] Agrinskaya N.V., Kozub V.I., Poloskin D.S. Semiconductors, **44**, 472 (2010)
[4] A. L. Efros, B. I. Shklovskii, *Electronic properties of doped semiconductors* (Springer-Verlag, 1984).
[5] Agrinskaya N.V., Kozub V.I., Poloskin D.V., Chernyaev A.V., Shamshur D.V. Jetp Lett., **80**, 30 (2004)

Angular and Temperature-related Specific Features of the Interaction between Light and Heavy Holes in Weakly Doped p-Ge:Ga at Low Temperatures

A.I. Veinger, A.G. Zabrodskii, T.V. Tisnek, and S.I. Goloshchapov

Ioffe Physico-Technical Institute of the Russian Academy of Sciences, Politekhnicheskaya 26, St.Petersburg, 194021, Russia

The microwave low temperature magnetoresistivity of the low doped (nondegenerated) *p-Ge* was investigated with the use of the electron spin resonance technique. This technique permits to detect the derivative of the microwave absorption with respect to magnetic field and its change with this field. In our case, this change is proportional to the magnetoconductivity of the sample under investigation. Because of the averaging time of the effective masses of the light and heavy holes is much more then microwave frequency (10 GHz), this method permits to investigate the reaction of the each kind of the holes on the magnetic field separately. The experimental results are compared with the theory of the classical magnetoresistivity effect.

Keywords: Doped semiconductors, magnetoresistance, hole effective masses, low temperatures, electron spin resonance.
PACS: 72.20.My

INTRODUCTION

The magnetoresistive effect belongs to classical kinetic effects and has been rather long studied [1, 2]. However, it is still difficult to distinguish between the contributions of each kind of carriers to the observed effect in materials with degenerate bands and, in particular, p-Ge. Changes in the mobility of each kind of holes in a magnetic field can be separately studied by using microwave frequencies [3–6]. As shown in these studies, the averaging of the hole effective mass in weakly doped p-Ge at low temperatures occurs during a time longer than 10^{-10} s, i.e., during the characteristic energy dissipation time. Thus, measuring the microwave magnetoabsorption (MMA) at a frequency of 10 GHz by using the technique of electronic spin resonance (ESR) in p-Ge makes it possible to separate the responses of light and heavy holes to a magnetic field.

The ESR technique records the derivative of the MMA with respect to the magnetic field (dP/dH) and, according to the properties of the effect, the dependence of this derivative on the magnetic field H must show a minimum when the following relation between the field and the mobility μ is observed:

$$\mu H \approx 1. \qquad (1)$$

Because of the presence of two kinds of holes in a p-Ge sample, the field dependence of the MMA must show two minima corresponding to the different mobilities of light and heavy holes [5]. Using relation (1), we can find, by analyzing the positions of these minima, the angular and temperature dependences of the mobilities of both light and heavy holes in p-Ge. In this communication, we present results of a study of this kind, carried out with weakly doped p-Ge samples rotated in two mutually perpendicular planes, $(1\bar{1}0)$ and $(11\bar{2})$. These results are compared with the existing classical theory of the magnetoresistive effect [2].

EXPERIMENTAL RESULTS

Microwave measurements were made on oriented p-Ge samples having the form of parallelepipeds with characteristic dimensions of $4\times1\times0.3$ mm^3, with a resistivity of about 10 Ω cm ($p \approx 3\times10^{14}$ cm^{-3}). The effect was observed at temperatures higher than 10 K, when strong thermal ionization of shallow Ga acceptors begins in Ge.

Figure 1 and 2 show experimental dependences of the derivative of MMA, (dP/dH), with respect to the magnetic field at several temperatures in relation to the angle φ between the magnetic field direction and the crystallographic axes. The angle φ was assumed to be zero when the second (strong-field) minimum related to heavy holes was found to be at the weakest field. It can be seen in the figures that the minima corresponding to light and heavy holes are, indeed, observed separately in the temperature range under study, as shown previously in [5]. In addition, it was found that the field dependences of the MMA are different when the samples are rotated in the $(11\bar{2})$ and $(1\bar{1}0)$ planes.

Upon a rotation by 90° in the $(1\bar{1}0)$ plane (Fig. 1), both the amplitude of the derivative of MMA and the position of the strong-field minimum of the derivative markedly change. At the same time, the position of the weak-field minimum remains nearly unchanged. It can be seen that, at high temperatures, the amplitude of the strong-field minimum decreases, whereas that of the weak-field minimum grows.

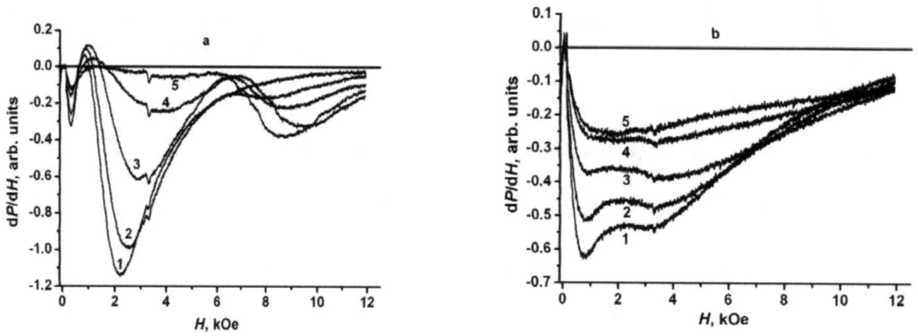

FIGURE 1. Field dependences of the magnetoabsorption at various angles between the magnetic field and the [110] direction in rotation in the $(1\bar{1}0)$plane: (1) 0°, (2) 20°, (3) 40°, (4) 60°, and (5) 80°. Temperature T, K: (a) 20 and (b) 80.

Figure 2 shows the same dependences for a sample rotated in the $(11\bar{2})$ plane at various temperatures. Comparison of Figs. 1 and 2 reveals the difference between the effects in the case of rotation in different planes. These are the substantially more pronounced changes in the position of the strong-field minimum corresponding to heavy holes upon rotation in the $(1\bar{1}0)$ plane, compared with that in the $(11\bar{2})$ plane. It should also be noted that, in the case of rotation in the $(1\bar{1}0)$ plane, the amplitude of the minimum related to heavy holes markedly decreases, whereas in rotation in the $(11\bar{2})$ plane, the decrease is on the order of 10%.

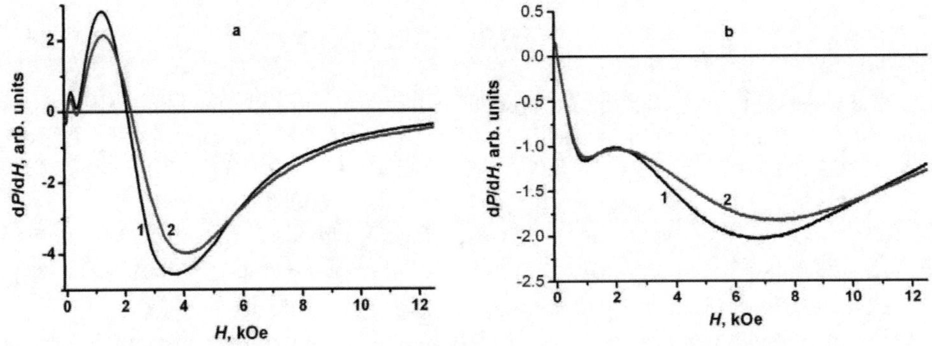

FIGURE 2. Field dependences of the MMA at various angles between the magnetic field and the [111] direction: (1) 0°, (2) 90° in rotation in the $(11\bar{2})$ plane. Temperature T, K: (a) 20 and (b) 80.

The difference between the field dependences of MMA in rotation in various planes is more visually illustrated by Fig. 3, which shows how the position of the minimum for heavy holes depends on the angle φ. The field at which the derivative of MMA is at a minimum depends not only on the rotation plane and angle φ, but also on temperature. At low temperatures, the fields of the minima become close at φ = 90°, whereas at high temperatures at angles φ ≈ 90° in rotation in the $(1\bar{1}0)$ plane, the field of minimum in the derivative of MMA is always substantially lower than that for rotation in the $(11\bar{2})$ plane.

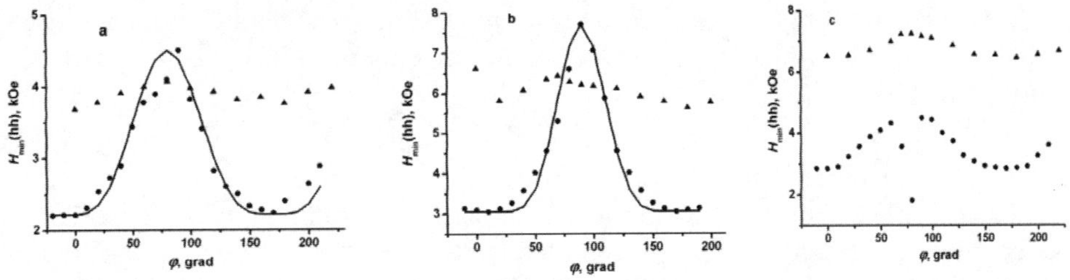

FIGURE 3. Dependence of the field of minimum in the derivative of MMA for heavy holes on the angle between the current flow direction and the [111] axis in the case of rotation in the $(11\bar{2})$ -(7) plane and between the current flow direction and the [001] axis in rotation in the $(1\bar{1}0)$ - (,)plane at various temperatures T, K: (a) 20, (b) 40, and (c) 60.

Journal of Physics: Conference Series **376** (2012) 012015 doi:10.1088/1742-6596/376/1/012015

It should be noted here that, in rotation about the $[1\bar{1}0]$ and $[11\bar{2}]$ axes, the extreme values of fields of minima in the derivative of MMA correspond to different directions in the sample. In the $(1\bar{1}0)$ plane, the minimum value of this field corresponds to the [001] direction, and the maximum value, to the [110] direction; in the $(11\bar{2})$ plane, the minimum and maximum fields correspond to the $[1\bar{1}0]$ and [111] directions, respectively. It follows from Figs. 1 and 2 that the position of the minimum related to light holes remains unchanged when the sample is rotated.

Thus, the angular and temperature dependences of MMA, related to light and heavy holes, are found to be rather complex.

DISCUSSION

The theory of the anisotropic magnetoresistive effect, which relates the effect to the anisotropy of hole effective masses, can be found in the monograph [2]. The effective mass is largest when the current is directed along one of the [001] axes. Along the [110] and [112] axes, the same intermediate value coinciding.

Let us consider how a change in the effective mass affects the measured MMA effect. The MMA is mostly affected by the transverse magnetoresistive effect. Therefore, if a sample is rotated in the $(1\bar{1}0)$ plane, the lowest value of the field of minimum must be observed when the field is directed along the [110] axis. In this case, the current making the largest contribution to the MMA flows along the [001] axis, in which direction the effective mass is the smallest. Upon a rotation by 90°, the [110] axis with a large effective mass becomes perpendicular to the field direction. Accordingly, the field at which the derivative is at a minimum shifts to stronger fields. This behavior can be seen in Figs. 1 and 3. However, the maximum field of minimum must be observed when the field is perpendicular to [111], the direction constituting an angle of about 55° with the [001] axis and an angle of about 35° with the [110] axis. The absence of a maximum of this kind indicates that the scattering anisotropy makes a significant contribution to the anisotropy of MMA.

When the sample is rotated in the $(11\bar{2})$ plane, the effective mass of heavy holes must change from values corresponding to the $[1\bar{1}0]$ current flow direction to those corresponding to the [111] current direction. In this case, the effective mass varies within a narrow range around its largest value. And just this behavior is observed in Figs. 2 and 3, in which the field of minimum remains high at any direction of the current in this plane. Thus, in the case of rotation about the mutually perpendicular $[1\bar{1}0]$ and $[11\bar{2}]$ axes, the extreme values of the minima in the derivative of MMA are observed at different directions in the sample.

The complex nature of the momentum scattering and the angular dependence of the effective mass at low temperatures hinder an exact analytical description of the behavior of the derivative of MMA. However, the experimental angular dependences make it possible to obtain, by means of fitting, rather adequate curves describing the low-temperature dependence of the minimum of the derivative of MMA in rotation in the $(1\bar{1}0)$ plane. It can be seen in Fig. 3 that the dependence must be sinusoidal with a period of 180°. With the designations $H_{[110]} = H_{min}$ and $H_{[001]} = H_{max}$, we, by analogy with [2], represent the angular dependence of the field of minimum, $H_{min}(\varphi)$, as

$$H_{min}(\varphi) = H_{min} + (H_{max} - H_{min}) \sin^n\varphi \, , \quad (2)$$

where the exponent n is the fitting parameter.

The calculated values of such dependences of the field for temperatures of 20 and 40 K are represented by the curves in Figs. 3a and 3b, respectively. It can be seen that these curves are in satisfactory agreement with the experimental points at an exponent $n = 4$ for $T = 20$ K and $n = 8$ for $T = 40$ K. The decrease in the exponent on lowering the temperature is in qualitative agreement with the fact that the time of momentum scattering is more strongly affected by the scattering on ionized impurities.

Let us consider the temperature dependences of the positions of minima in the derivative of MMA in the $(1\bar{1}0)$ and $(11\bar{2})$ planes. The dependences are shown in Fig. 4 in the log–log scale. It can be seen that, in accordance with the power-law temperature dependences of the mobility, the temperature dependences of the minimum in the derivative of MMA, indeed, have the form of straight line portions. There are, however, a number of specific features that make it possible to judge about interaction modification of light and heavy holes with temperature.

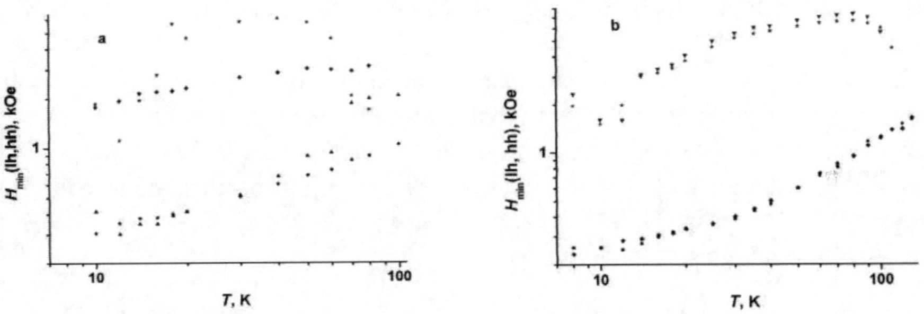

FIGURE 4. Temperature dependences of the minimum in the derivative of MMA for light and heavy holes at various angles between the magnetic field H and the [110] axis. (A) Rotation in the $(1\bar{1}0)$ plane: light holes: , - 0° and (7) 90°; heavy holes: (Λ) 0° and (ξ) 90°. (B) Rotation in the $(11\bar{2})$ plane: light holes: (,) 0° and (B) 0°; heavy holes: (Λ) 0° and (7) 90°.

It should be noted in the first place that, for heavy holes and the current flowing in the $(1\bar{1}0)$ plane in the [110] direction (magnetic field is directed at an angle of 90° to the [110] axis), the dependence differs from that of a power-law type. In this case, the position of the minimum strongly depends on temperature. This indicates that, for the current flowing in this direction, the scattering mechanisms at low and high temperatures are different.

The rest of the dependences (for light and heavy holes in other directions and planes) are, as noted above, of the power-law type. At temperatures lower than 30 K, they are proportional to $T^{1/2}$, which points to the prevalent scattering on acoustic phonons. In this temperature range, intersubband transitions little affect the mobility of heavy and light holes, and the temperature dependences in Fig. 4 are nearly parallel.

However, at temperatures higher than 30 K, the curves for light and heavy holes start to gradually become closer, which reflects the enhancement of scattering between the subbands of light and heavy holes.

In other words, there occurs stronger mixing of light and heavy holes. As a result, the average mass of light holes increases and that of heavy holes decreases. At high temperatures the field of minimum in the derivative of MMA even decreases. The strong-field minimum approaches the weak-field minimum because the effective masses of heavy and light holes are averaged during a time that is substantially shorter than 10^{-10} s, the period of the microwave oscillations we used. This corresponds to the well-known fact: at high temperatures the mobility in the hole system of p-Ge, manifested in the magnetoresistive effect is close to that of light holes.

CONCLUSIONS

A study of the angular and temperature dependences of the microwave magnetic-field-dependent absorption in lightly doped p-Ge revealed a number of interesting specific features associated with the band degeneracy.

1. The anisotropy of the MMA by holes in a magnetic field is determined by the anisotropy of their mobility.
2. The anisotropy of heavy holes mobility at low temperatures ($10K \leq T \leq 30$ K) is determined both by the anisotropy of the effective mass and by the anisotropy of momentum scattering.
3. The averaging time of the effective masses of light and heavy holes in the temperature range $10 \text{ K} \leq T \leq 30$ K substantially exceeds 10^{-10} s. Therefore, it is not manifested in the field and temperature dependences of the MMA.
4. The interparticle interaction becomes more pronounced with increasing temperature. The effect of the interaction between holes gradually becomes noticeable at temperatures higher than 30 K.

ACKNOWLEDGMENTS

The study was financially supported by the Russian Foundation for Basic Research (grant no. 10-02-00629), Ministry of Education and Science of the Russian Federation (RF Presidential grant NSh-3306.2010.2), Presidium and Department of physical sciences of the Russian Academy of Sciences.

REFERENCES

1. K. Seeger Semiconductor Physics. Springer-Verlag, Wien, NY, 1973.
2. A.C.Beer. Galvanomagnetic effects in semiconductors. Acad.Press, NY, L 1963
3. A.I. Veinger, A.G. Zabrodskii, T.V. Tisnek, S.I. Goloshchapov. Sol. St. Comm. **133**, 455, (2005).
4. A.I. Veinger, A.G. Zabrodskii, T.V. Tisnek, S.I. Goloshchapov Phys. stat. sol. (c) **3**, 317, (2006).
5. A.I. Veinger, A.G. Zabrodskii, T.V. Tisnek, S.I. Goloshchapov Phys. stat. sol. (c) **5**, 835, (2008).
6. A.I. Veinger, A.G. Zabrodskii, T.V. Tisnek. Semicond, **36**, 772, (2002).

14th International Conference on Transport in Interacting Disordered Systems (TIDS-14) IOP Publishing
Journal of Physics: Conference Series **376** (2012) 012016 doi:10.1088/1742-6596/376/1/012016

Magnetoresistance in two-dimensional array of Ge/Si quantum dots

N. P. Stepina, E. S. Koptev, A. G. Pogosov, A. V. Dvurechenskii, A. I. Nikiforov, E.Yu. Zhdanov

E-mail: stepina@isp.nsc.ru
Institute of Semiconductor Physics, 630090 Novosibirsk, Russia

Abstract. Magnetoresistance in two-dimensional array of Ge/Si was studied for a wide range of the conductance, where the transport regime changes from hopping to diffusive one. The behavior of magnetoresistance is similar for all samples; it is negative in weak fields and becomes positive with increasing of magnetic field. Negative magnetoresistance can be described in the frame of weak localization approach with suggestion that quantum interference contribution to the conductance is restricted not only by the phase breaking length but also by the localization length.

1. Introduction

Our previous study of temperature dependence of conductance and its non-linearity in 2D Ge/Si quantum dot (QD) array [1] showed that change of the structural parameters of the system results in the transition from strong (SL) to weak localization (WL) behavior and, respectively, from hopping to diffusive transport. Different transport regimes should be characterized by different behavior of magnetoresistance (MR). The classical theory for hopping MR predicts a positive MR due to the shrinkage of the impurity wave functions in magnetic field which leads to the decrease of the overlap between localized states and thus reduces the hopping probability [2]. Negative MR is well known at WL [3], where the interference among multiple elastic scattering paths of a single conducting electron leads to the enhancement of the backscattering probabilities. A phase shift in the electron wave function introduced by magnetic field suppresses the coherent backscattering and produces the negative MR. Negative MR was also found for variable-range hopping (VRH) [4], but the mechanism of this effect has different nature. In VRH interference the backscattering gives a negligible contribution to negative MR and the major source of the effect is destructive interference between the forward paths [5].

In this paper we describe the results of MR measurements performed on 2D Ge/Si QD system in which the localization strength can be adjusted by variation of QD occupancy, QD array density and under the annealing at 480÷625°C. For the samples under study the range of the conductances at 4.2 K is approximately $10^{-5}\div10^{-10}$ Ohm^{-1}.

2. Experimental setup

The samples were grown on a (001) p-Si substrate with a resistivity of 20 Ω·cm by molecular-beam epitaxy of Ge in the Stranskii-Krastanov growth mode. Two growth regimes allowed to obtain the QD arrays with different density. In the first case, the growth temperature for 10 monolayers (ML) Ge layer was 300°C and the growth rate was 0.2 ML/s. As a results, the areal density of the dots was shown to be $\sim 4 \times 10^{11}$ cm^{-2}. We named this type of samples as single density samples (SD). In the second case, the decrease of the Ge growth temperature down to 275°C with simultaneous increase of the growth rate allowed to reach the twice higher QD areal density ($\sim 1 \times 10^{12}$cm^{-2}). These samples were referred to as double density samples (DD). To supply holes to the dots, a boron δ-doped Si layer was inserted 5 nm below the Ge QD layer. The SD samples were additionally annealed in Ar atmosphere during 30 minutes at 550, 575, 600 and 625 °C. The silicon cap layer had a thickness of 40 nm. Al metal source and drain electrodes were deposited on the top of structure and heated at 480°C to form reproducible Ohmic contacts. The resistance along the QD layer was measured by the four-terminal method. The temperature stability was controlled using Ge thermometer. The magneto-transport measurements were carried out at 4.2 K in magnetic fields 0÷10 T.

3. Results and discussions

3.1. Temperature dependence of conductance

Fig 1a. demonstrates the temperature dependence of sheet conductance as Arrhenius plots for the following samples: for SD samples with filling factor ν ≈2.4 annealed at different temperatures (number 2-6), for SD sample with ν ≈2 annealed at 550°C (7) and for the sample with double density of dots (ν ≈3) in zero magnetic field (1). One can see that the conductance strongly depends on sample parameters and changes by the several order of magnitude. Annealing at 480÷625°C leads to the increase of the conductance with annealing temperature. The G values of DD sample (1) and SD samples (2, 3) annealed at high temperature lie in the range typical for the diffusive regime [6].

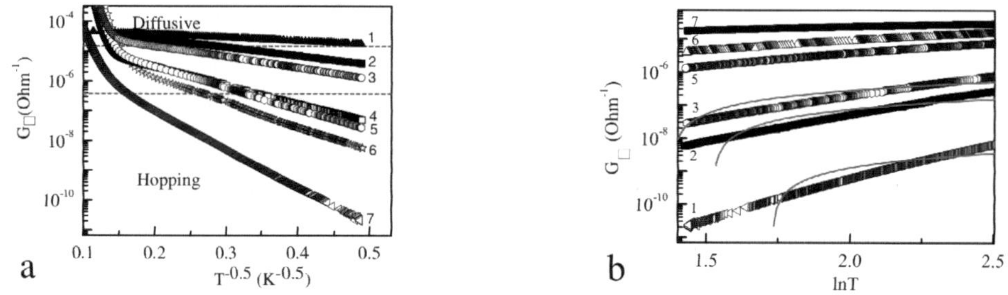

Figure 1. Temperature dependence of sheet conductance for the samples with different structural parameters. 1-DD sample with $\nu \sim 3$, 2-6: SD samples with $\nu \sim$2.4 and different temperature of annealing, °C: 2-625, 3-600, 4-575, 5-550, 6-480; 7-SD sample with $\nu \sim$2, 480°C; a - $G(T^{-0.5})$ dependence, b - approximation of $G(\ln(T))$ in the frame of WL approach.

The transport regimes for the samples under study were analyzed in the frame of both VRH and WL approaches. In general, the temperature dependence of conductance for VRH is given by

$$G(T) = \gamma T^m exp[-(T_0/T)^x] \quad (1)$$

where γ and m are the constants, T_0 is the material-dependent constant, $x = 1/3$ (Mott law) and $x = 1/2$ (Efros-Shklovskii (ES) law) for the 2D VRH without and with electron-electron interaction, correspondingly. To determine the behavior of $G(T)$, we analyzed the temperature dependence of the reduced activation energy $w(T) = \partial \ln G(T)/\partial \ln T = m + x(T_0/T)^x$ using the

Journal of Physics: Conference Series **376** (2012) 012016 doi:10.1088/1742-6596/376/1/012016

method proposed in [7]. The analysis shows that the temperature dependences for the samples 4-7 are well described by the ES law with $x \approx 0.5$. Localization length ξ can be determined from the T_0 value as $\xi = Ce^2/\varepsilon k_B T_0$, where theoretical value of constant C for single-electron hopping in 2D is $C = 6.2$ [8], and ε is the static dielectric response, k_B is the Boltzmann's constant. It was found that the $G(T)$ dependence for the samples with $G \geq 10^{-6}$ Ohm^{-1} can be described by the same equation but with poor accuracy. Nevertheless, the same procedure was carried out for all samples and both T_0 and localization length ξ were determined. The values of ξ for each sample, calculated using the above mentioned formula, are marked in Table 1. One can see that the smaller is the conductance the more localized is the system as it should be in accordance with the classical scaling theory [9]. For the samples 1, 2 and 3 (most conductive ones) the formally determined values of ξ are about of several micrometers. Of course, such a large ξ magnitude, poor accuracy in the evaluation of $G(T)$ by the VRH, high value of conductance ($\sim e^2/h$) and weak temperature dependence of conductance suggest that the theory of quantum correction should be applied for describing the transport behavior in this kind of samples. Fig.1b demonstrates the approximation of all samples under study (the same number of samples that for Fig.1a) in a frame of the theory of quantum corrections (the G axis is shown in logarithmic scale for better vision of all experimental curves). One can see that the temperature dependence of conductance is well described by the logarithmic law (gray lines) only for samples with $G \geq 10^{-6}$ Ohm^{-1}. Our previous results [1] showed that at $G \leq 10^{-2}e^2/h$ the conductance of the system is well described by the variable-range hopping via Coulomb gap states (Efros-Shklovskii (ES) law), whereas at $G \geq 0.4e^2/h$ the conductance is diffusive. Corresponding bounds are shown in Fig.1a as dash lines. Thus, the analysis of the temperature dependences of conductance for samples under study agrees with the G- bounds for the transition from hopping to diffusive transport observed in [1].

3.2. Magnetoresistance

The typical transversal field sweeps of R/R_0 at fixed temperature (4.2 K) are shown in Fig.2. One can see that all samples demonstrate similar behavior of MR in spite of the differences

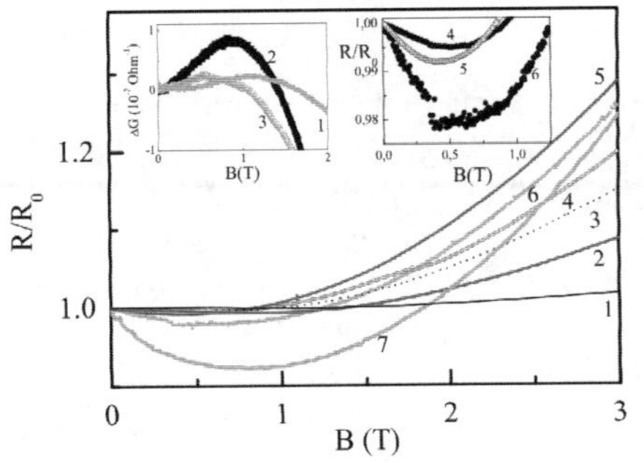

Figure 2. Magnetoresistance for the same samples that in Fig.1. Insets- enlarged images for high-conductance samples (left inset) and for low-conductance samples(right inset).

in the conductance and transport behavior. In low fields one can observe the negative MR that turns into positive one with the increase of magnetic field. To understand what kind of mechanism can be responsible for the negative MR in our system, we have analyzed all of our MR data in the frame of both VRH and WL approach. Insets to Fig.2 show the enlarged images of low-field data for high-conductance samples in $\Delta G(B)$ axis (left inset) and for low-conductance samples in $R/R_0(B)$ coordinates (right inset). In the frame of NSS model for the VRH regime [4], averaging over different realization of hopping paths of carriers scattering on N centers with $\mu = 8\pi\xi\epsilon_i/(\epsilon_1 - \epsilon_i) < 0$ scattering amplitude leads to the linear dependence of

$L(B,r) \equiv \ln(\sigma(B)/\sigma(0))$ on magnetic field:

$$L(B,r) \approx N\mu^2\xi(|\mu|\,\xi)^{1/2}rB \tag{2}$$

where r is the hopping length, ϵ_1 is the energy of tunneling electron, ϵ_i is the energy of scatterer. As for VRH via Coulomb gap states $r \approx \xi(T_0/T)^{1/2} \approx \xi^{1/2}$, we can rewrite the expression (2): $L \approx C\xi^{4.5}$, where C does not depend on ξ. Thus, for larger ξ the larger the effect of negative MR should be. The linear dependences of $\ln(R/R_0)$ on B are really observed practically for all low-conductive samples (Fig.3), but the effect of negative MR falls with the ξ increase (inset to Fig.3). It means that the nature of negative MR differs from the interference of forward paths in VRH.

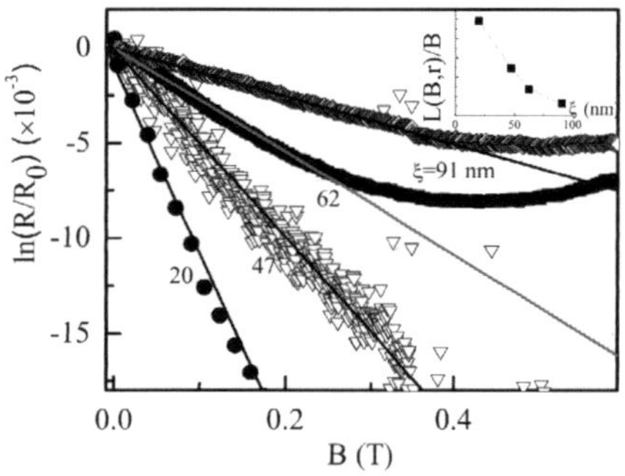

Figure 3. MR effect for low-conductive samples. Inset- slope of the linear approximation of the experimental data versus ξ.

All the samples under study were analyzed in the frame of WL approach. In accordance to the phase breaking time (τ_ϕ) approximation [10,11] due to the ordinary quantum correction to the 2D system conductivity, the conductance changes with magnetic field as (HLN expression)

$$\delta G(B) = \alpha\frac{e^2}{2\pi^2\hbar}\left[\psi(\frac{1}{2}+\frac{\hbar}{2Be(L_\phi)^2}) - \psi(\frac{1}{2}+\frac{\hbar}{2Bel^2}) + 2\ln(\frac{L_\phi}{l})\right], \tag{3}$$

where l is the transport mean free path, α is the constant of the order of unity, $\psi(x)$ is digamma function and the phase breaking length is $L_\phi = (2D\tau_\phi)^{1/2}$, where D is the diffusion coefficient. Although the prefactor α has to be equal to unity within the framework of the conventional WL theory, it is often used as the second fitting parameter together with L_ϕ. An important point is that almost all experimental data are better fitted with $\alpha \leq 1$ [12], which contradicts to the theory. Especially α vanishes when k_Fl becomes close to unity. In this case to fit experimental data with HLN expression, one needs to vary 3 parameter: L_ϕ, l and α. To resolve this problem, we use the method of successive approximation [13] when solving together the HLN equation and the equation for the conductance in WL regime. As far as the correction due to electron-electron interaction is small at $k_Fl \approx 1$, the only correction due to interference was taken into account:

$$G(T) = G_{dr} - 2G_0 \ln\left(\frac{L_\phi(T)}{l}\right) \tag{4}$$

The l value can be extracted from the value of Drude conductance G_{dr}: $G_{dr} = e^2n\tau/m = \pi k_F l G_0$ here $G_0 = e^2/2\pi^2\hbar$, n and m denote electron density and mass, respectively, τ is the elastic transport mean free time, $\nu = m/2\pi\hbar^2$.

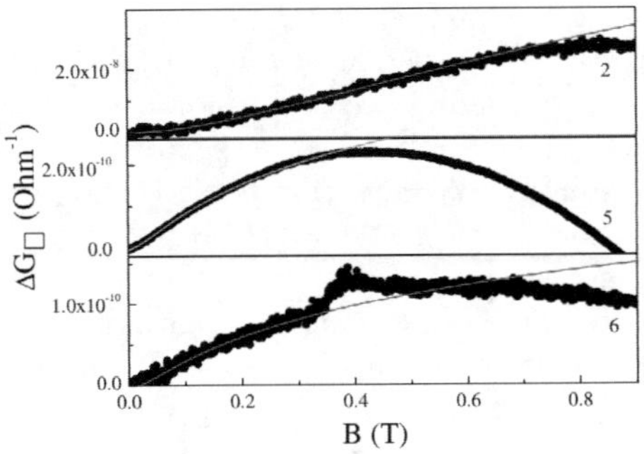

Figure 4. Approximation of the experimental data by HLN equation for samples 2, 5 and 6 from Fig.1.

The example of the experimental data approximations with HLN equation for 3 samples: 2, 5 and 6 are shown in Fig.4. The parameters α, L_ϕ^* and l were found from 10-12 successive approximations (the asterik in L_ϕ^* is because it is unknown in advance how the fitting parameter in the HLN expression relates to the true phase breaking length at low conductances). It is necessary to note that decrease of the conductance leads to a strong decrease of α value. The authors [14] maintain that it is a typical behavior of α when the system is in the crossover between WL and so called weak insulator (WI) regime. The last one is characterized by the zero-B conductivity that can be less than G_0 at low T and $k_F l \geq 1$. Namely this regime is supposed to observe in our case. Generally speaking, at $k_F l \geq 1$, the nature of interacting electron transport in the case when the states are localized with large enough localization length ξ is not fully understood. The authors [15] showed that the MR even at low conductance, $G(B_0) \leq G_0$, can be still fitted by the HLN formula derived for $k_F l >> 1$, but with a reduced prefactor $\alpha << 1$. However, they examine the regimes with G not smaller than $10^{-2}G_0$. We carried out this procedure for all samples under study up to $G \sim 10^{-4}G_0$ and the resulting parameters are collected in Table 1.

Table 1. Characteristic lengths, nm

	l	L_ϕ^*	ξ^*	$k_F l$	ξ	α
6	14	134	135	1.43	47	6×10^{-6}
5	14	142	141	1.45	62	1×10^{-5}
4	12	86	85	1.23	91	3×10^{-5}
2	11	51	59	1.08	390	5×10^{-3}
1	5	12	24	1.04	2250	0.79

As it was expected, $k_F l$ values for the samples under study lie in the range 1-1.5. The typical distance between quantum dots is about 5 nm for sample with the dot density $\approx 10^{12}$ cm^{-2} and 15 nm for the samples with $\approx 4\times10^{11}$ cm^{-2} that are close to the obtained l values for the samples without high-temperature annealing. If we suggest that the main scattering centers for the delocalized electrons are the QDs, the annealing should lead to the smearing of dots and, consequently, to the decrease of the inter-dot distance. Decrease in l value with annealing confirms this proposal and allows one to represent Ge nanoclusters as scattering centers. The characteristic values of L_ϕ^* lie in the range 15 - 140 nm. In the frame of proposed approach

[15], L_ϕ^* value is the combination between dephasing length and localization radius and we cannot extract one of these lengths separately. Using the expression for $\xi^* \approx l\exp(\pi k_F l/2)$, we can estimate the value ξ^*. The result of these calculations are added to Table 1. One of the intriguing result is that L_ϕ^* and ξ^* have equal values for all samples except the most conductive one. This coincidence can be understood assuming that the interference effects happen inside the cluster with characteristic size ξ^*. In this case, the interference correction is restricted not only by phase breaking length, but also by localization length ξ^*: $1/(L_\phi^*)^2 = 1/(L_\phi)^2 + 1/(\xi^*)^2$. For $L_\phi >> \xi^*$, ξ^* plays the role of the sample dimensions when considering the quantum coherence corrections to the conductivity. On a shorter scales, the electron dynamics remain diffusive: an electron merely 'does not know' that it is going to be localized. In this case the value L_ϕ in HLN equation should be replaced with ξ^*, and this explains the observed coincidence of L_ϕ^* and ξ^*. When $L_\phi \approx \xi^*$, $L_\phi^* = \xi^*/\sqrt{2}$. In our case for sample 1 we obtain $L_\phi^* = \xi^*/2$ than the relation between L_ϕ and ξ^* is: $L_\phi = \xi^*/\sqrt{3}$. The difference in ξ^* and ξ obtained from the analysis of $G(T)$ dependence is evident, because the first length is determined by the interference inside the cluster with ξ^*- size and the second one defines the motion of this cluster along the hopping network. To understand if the proposed model is valid or not, the temperature dependences of MR need to be known.

4. Acknowledgments

This work is done with support of RFBR (Grant 10-02-0618). Authors are thankful to G.M.Minkov and A.V. Nenashev for very useful discussions.

5. References

[1] N.P. Stepina, E.S. Koptev, A.V. Dvurechenskii, A.I. Nikiforov 2010 *Phys. Rev.* **B80** 23
[2] B.I. Shklovskii 1983 *Fiz. Tech. Poluprovod.* **17** 2055
[3] G. Bergmann 1984 *Phys.Rep.* **1** 107
[4] I.E. Shlimak and A.N. Ionov 1991 in *Hopping Transport in Solids* edited by M.Pollak and B.I. Shklovskii
[5] V.L. Nguen, B.Z. Spivak, and B.I. Shklovskii 1985, *Zh.Eksp. Teor. Fiz.* **89** 1770
[6] V.F. Gantmakher 2003 in *Electrons in non-ordered medium*, Fizmatlit, Moscow
[7] A.G. Zabrodskii and K.N. Zinoveva 1984 *Zh. Eksp. Theor. Fiz.* **86** 727 [*Sov. Phys. JETP.* 1984 **59** 425]
[8] D.N. Tsigankov and A.L. Efros 2002 *Phys. Rev. Lett.* **88** 176602
[9] E. Abrahams, P.W. Anderson, D.C. Licciardello, T.V. Ramakrishnan 1979 *Phys. Rev. Lett.* **42** 673
[10] B. L. Altshuler, D. E. Khmelnitsky, A. I. Larkin, and P. A. Lee 1980 *Phys.Rev.* **B22** 5142
[11] S. Hikami, A. Larkin, and Y. Nagaoka 1980 *Prog. Theor. Phys.* **63** 707
[12] G.M. Minkov, O.E. Rut, A.V. Germanenko, A.A. Sherstobitov, V.I. Shashkin, O.I. Khrykin, and B.N. Zvonkov 2003 *Phys. Rev.* **B 67** 205306
[13] G.M. Minkov, O.E. Rut, A.V. Germanenko, A.A. Sherstobitov, B.N. Zvonkov, E.A. Uskova, and A.A. Birukov 2002 *Phys. Rev.* **B65** 235322
[14] G.M. Minkov and A.V. Germanenko, I.V. Gornyi *Phys.Rev.* 2004 **B 70** 245423
[15] G. M. Minkov, A. V. Germanenko, O. E. Rut, A. A. Sherstobitov, and B. N. Zvonkov 2007 *Phys.Rev.* **75** 235316

Journal of Physics: Conference Series **376** (2012) 012017 doi:10.1088/1742-6596/376/1/012017

Interference mechanism of magnetoresistance in variable range hopping: the effect of paramagnetic electron spins.

A.V. Shumilin and V.I. Kozub
Ioffe Physical-Technical Institute of the Russian Academy of Sciences, 194021, Saint Petersburg, Russia.
E-mail: hegny@list.ru

Abstract.
The work is aimed to modify the conventional approach to interference magnetoresistance for the situation when there is an admixture of free spins. We considered the system with arbitrary probability P_{free} of electron spin to be free. We studied analytically and numerically both negative interference magnetoresistance and positive magnetoresistance due to spin ordering in magnetic field. We obtained the expression for magnetoresistance for two limiting cases corresponding to small and large number of scatterers in the hopping act. For both cases we obtained the explicit expressions for the dependence of the magnetoresistance on temperature and P_{free}. Our results are in semiquantitative agreement with experiment.

1. Introduction

At low temperature the conductivity in semiconductors with compensation is supported by hopping of the carriers between the localized states on the impurities. With low enough temperature the characteristic act of hopping occurs not between the neighboring impurities but between the closest impurities lying in the thin energy strip around the Fermi level. This phenomena is known as variable range hopping conductivity [1].

It is known that at weak magnetic fields ($H \leq 1\,\mathrm{T}$) the semiconductors in variable range hopping regime display negative magnetoresistance that appears to be linear in the magnetic field. This phenomena was first understood by Nguen, Shklovskii and Spivak [2] (for detailed review see [3]). It is related to the interference. During the hop electron scatters on the impurities that are outside of the conduction strip. So the resulting hopping amplitude is the sum of different tunneling paths, which interfere among themselves. The nature of hopping conductivity is such that destructive interference is strongly emphasized and thus the interference contribution leads to a decrease of the conductivity. The magnetic field suppresses the interference and leads to the negative magnetoresistance.

The theory of interference magnetoresistance [3] was based on the model of spinless electrons. This model can be justified when all electron spins are frozen, for example, by the exchange interaction. In [3] it was stated that there is no interference effect on the magnetoresistance when all electron spins are free. However this statement lacks the solid theoretical proof. Also there were no discussion of the system in which some spins are frozen and some are free. So

Journal of Physics: Conference Series **376** (2012) 012017 doi:10.1088/1742-6596/376/1/012017

there remains a question "how many free spins one should have to suppress the interference magnetoresistance".

Then in [4] it was also argued that the admixture of free spins leads to additional positive magnetoresistance due to spin ordering in the magnetic field. However no detailed consideration of this mechanism of magnetoresistance was made. For example, its detailed temperature dependence remains unknown.

There are two different limits for the theory of interference magnetoresistance. In [2, 3] the system with large number of scatterers in the hopping act was considered. In our work we call such a situation "the case of long hops". For this model only qualitative results were obtained in [2, 3]. Later it was shown in [5, 6] that in many occasions the characteristic number of scatterers is small. Most hopping acts occur without scattering, and to consider interference magnetoresistance one can take into account only hopping acts with a single scatterer. We will call this situation "the case of short hops". For short hops and spinless electron problem [5, 6] gives quantitative results for magnetoresistance.

There were several attempts [7, 8, 9] to make a quantitative theory for long hops. However, to our opinion, all these works do not pay sufficient attention to the dispersion of scatter energies. We will discuss this detail later. We also note that the existing theory of interference contribution to magnetoresistance is not sufficient to describe existing experimental data. For example there were observations of the suppression of negative magnetoresistance [10, 11] at low temperature that can not be described with conventional theory.

The goal of the present study is to develop a theory of interference contribution to magnetoresistance that explicitly takes into account the admixture of free spins. We consider two general problems. The first is to what extent the interference magnetoresistance is suppressed when some spins are free. The second is to study the spin ordering magnetoresistance.

The important feature for our consideration is the spin structure of frozen spins in impurity band. We assume that frozen spins form the so called Bhatt-Lee phase [12].

The common way to treat hopping conductivity is to introduce the Miller and Abrahams network of effective resistors. Within the framework of such an approach one calculates the mean hopping rate between the pairs of hopping impurities Γ_{ij}. The effective conductance between this impurities is $R_{ij} = T/e^2\Gamma_{ij} \propto \exp \xi_{ij}$. Here ξ_{ij} is the logarithm of resistance. Then with percolation theory one can estimate the critical (percolation) value ξ_c. The resistivity of the whole system is then proportional to $\propto \exp \xi_c$. The magnetoresistance can be calculated as (see [3])

$$\ln \frac{R(H)}{R(0)} = -\left\langle \ln \frac{\Gamma_{ij}(H)}{\Gamma_{ij}(0)} \right\rangle . \tag{1}$$

Here H is the magnetic field. The averaging is taken over pairs of impurities that correspond to critical resistors (i.e. with $\xi_{ij} = \xi_c$).

2. Short hops

Let us first discuss the situation when the characteristic number of scatterers in a hopping act is small. The interference magnetoresistance in this case is controlled by the hopping resistors with one intermediate impurity. As it was shown in [6], this situation (of "short hops") corresponds to many typical experimental realizations. The opposite situation is the case of "long hops" when the characteristic hopping resistor contains many scattering impurities.

In the case of short hops each resistor of interest has either a scatterer with an electron with frozen spin, a free scatterer without an electron or a scatterer with a free electron spin. The contributions of these resistors are additive (1). The first two types of resistors can be described with the spinless electron theory. For short hops this theory is a rigorous one (see [5, 6]). Its has a following result

$$\ln \frac{R(H)}{R(0)} \propto -r_h^{2+d/2} H, \tag{2}$$

where r_h is the mean hopping distance and d is the system dimensionality. The dependence on r_h controls the temperature dependence of magnetoresistance as $r_h \propto T^{-1/(d+1)}$ for the Mott conductivity and $r_h \propto T^{-1/2}$ for Efros-Shklovskii conductivity over Coulomb gap states.

So we have to estimate only the contribution of the hopping resistors with free electron spins. For this resistors it is important to note that Γ_{ij} corresponds to time average of the hopping acts. To calculate it one should take into account that sometimes the tunneling electron has the same direction as the spin on the scatterer and sometimes their directions are opposite. There are always two tunneling paths in this system: one that includes the scatterer and one that does not contain it. If the spins of the sites involved have the same direction, the corresponding paths interfere, otherwise they do not interfere. So we can write the following expression for Γ_{ij} for free intermediate spin

$$\Gamma_{ij} \propto P_{\uparrow\uparrow} \left| J_1 + J_2 e^{i\varphi} \right|^2 + (1 - P_{\uparrow\uparrow}) \left(|J_1|^2 + |J_2|^2 \right). \tag{3}$$

Here $P_{\uparrow\uparrow}$ is the probability for the two spins to have the same direction. It depends on the magnetic field as the magnetic field aligns free spins. J_1 and J_2 are, correspondingly, the amplitudes of tunneling path that does not contain the intermediate impurity and the one that includes it. $\varphi = HS/\Phi_0$ is the phase difference between this pathes. S is the area between the pathes and Φ_0 is the magnetic flux quantum.

To get the expression for magnetoresistance one should substitute (3) into (1) and perform the averaging. The resulting magnetoresistance is quadratic on H. There are many magnetoresistance mechanisms that give the quadratic field dependence (not necessary related to interference) so we keep only one term in this magnetoresistance that has unusually strong temperature dependence and thus can be discriminated from other mechanisms (for example from wavefunction shrinkage magnetoresistance). It is the spin ordering term

$$\propto P_{free} n r_h (r_h a)^{(d-1)/2} \left(\frac{\mu_b g H}{T} \right)^2. \tag{4}$$

Here n is the impurity concentration. P_{free} is the relative part of the scattering impurities that has an electron with free spin. For three-dimension system with Mott conductivity this term depends on temperature as $\propto T^{-5/2}$. To compare the temperature dependance of wavefunction shrinkage magnetoresistance is $\propto T^{-3/4}$.

To summarize the results of this section let us combine the different contributions to magnetoresistance:

$$\ln \frac{R(H)}{R(0)} \propto -C_1 (1 - P_{free}) r_h^{2+d/2} H + C_2 P_{free} n r_h (r_h a)^{(d-1)/2} \left(\frac{\mu_b g H}{T} \right)^2, \tag{5}$$

where C_1 and C_2 are constants independant on temperature and magnetic field.

3. Long hops.

Let us now discuss the case of long hops. In this case there is an exponentially large number of tunneling paths (even without backscattering), 2^N, where N is the number of scatterers.

There where previous studies of interference magnetoresistance in the long hops case at least for spinless electrons (see e.g. [7, 8, 9]). However to our opinion these studies underestimate the dispersion of the tunneling path amplitudes J_i. Some of the works adopt the model when there is no dispersion of $|J_i|$ (the tunneling path amplitudes differs only with sign). Some adopts the model when the distribution of J_i is the one with zero mean value and the dispersion of normalized tunneling amplitudes of the order of unity.

However we argue that the tunneling path amplitudes actually has an exponentially broad log-normal distribution. The amplitude of some tunneling path i can be expressed as

$$J_i = I_{0,k_1} \prod_{k(i)}^{N_i+1} \frac{I_{k-1,k}}{E_k}. \quad (6)$$

Here the index $k(i)$ enumerates the scattering impurities that participate in the tunneling path i. N_i is a number of this impurities. Usually $N_i \sim N/2$, where N is the total number of scatterers in the resistor.

So the path amplitude is the product of a large number of random terms. Due to the central limit theorem it is distributed with a log-normal law with dispersion that increases exponentially with N. Actually it is an exponentially broad distribution. This fact can be also confirmed with computations (see fig. 1 (a)).

One of the properties of the log-normal distribution is the fact that the mean value is exponentially larger then the value corresponding to the maximum of distribution density. So the sum of a large number of these paths (for example the net tunneling amplitude that controls Γ_{ij}) is controlled by a small number of the summands. We call the paths corresponding to these summands as "significant paths". On fig. 1 we show that the relative number of this paths is exponentially small.

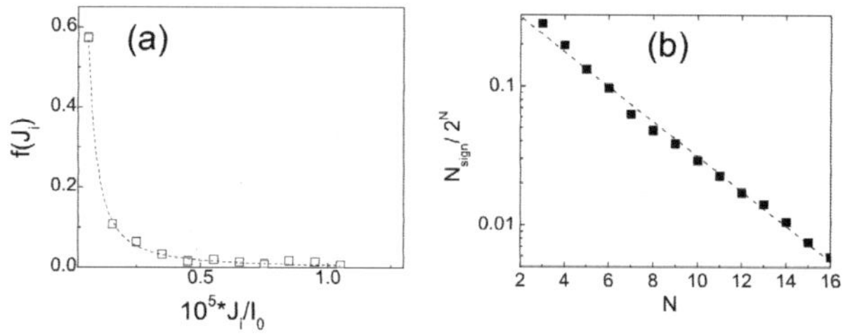

Figure 1. (a) — distribution of absolute values of tunneling path amplitudes J_i. Points are computed values, line is the approximation with log-normal law. (b) — the relative part of significant path versus scatterers number.

To understand the interference effects in the case of long hops one should understand the nature of the significant paths. The numerical computations show that these paths are similar to each other in some way. This fact is related to the dispersion of spatial locations and of energies of the scatterers. One can see from (6) that the inclusion of some scatterer j into the tunneling path is equivalent to the multiplication of path amplitude by the factor

$$\mu_j = \frac{I_{j-,j} I_{j,j+}}{I_{j-,j+} E_j}. \quad (7)$$

Here $j-$ is the previous (to $j-th$) scatterer included into the path under consideration. $j+$ is the closest scatterer after $j-th$ included into the path. I_{ij} is the energy overlap integral. E_j is the energy of scatterer j measured from the Fermi level.

Strictly speaking μ_j depends not only on the characteristics of scatterer j but also on the rest of the tunneling path. However basing on the typical values of energy and position of the scatterer one can estimate μ_j for a characteristic path. Let the overlap integrals have the form

$I_{ij} = I_0 \exp(-r_{ij}/a)$. Also we assume that all considered scattering impurities lie in a thin strip around direct tunneling line. Then μ_j for the characteristic path can be estimated as

$$\mu_j = \exp\left(-\frac{\rho_j^2}{a r_{nn}}\right)\frac{I_0}{E_j}. \tag{8}$$

Here ρ_j is the distance of scatterer j from the tunneling line. r_{nn} is the typical distance between impurities included into the tunneling path.

With above arguments all scatterers can be separated into three groups. First there are scatterers with $|\mu_j| \ll 1$. These scatterers are not included into the significant paths (at least into most of them) as inclusion of this impurities make tunneling path amplitude smaller. All impurities that are far from tunneling line fall into this group. In our work we call this impurities irrelevant. They can easily be excluded from consideration. Then there are impurities with $\mu_j \gg 1$. Such impurities are included into most significant pathes because they significantly increase the path amplitude. We call these impurities the backbone ones. Finally there are impurities with $\mu_j \sim 1$ that do not fall into the previous groups. We call them the interference impurities.

We made numerical computations to find whether the backbone impurities really exist in the system. We considered the distribution density of impurity energy to be constant within the interval $-I_0 < E_j < I_0$. The computations show that the backbone impurities exist in the system. Their number is comparable to the number of interference scatterers and their ratio seems to be independent on hopping distance and impurity concentration.

To go further we adopt the simple model that is valid when any two interference scatterers are separated from each other by at least one backbone impurity. In this case one can express the typical significant path amplitude as a product of factors corresponding to different scatterers. With this expansion one can get the following expression for magnetoresistance in the long hop limit

$$\ln\frac{R(H)}{R(0)} \propto (n r_h \rho_{max}^{d-1})\left(-C_1\left(1-P_{free}\right) r_b^{2+d/2} H + C_2 P_{free} n r_b (r_b a)^{(d-1)/2}\left(\frac{\mu_b g H}{T}\right)^2\right). \tag{9}$$

Here $\rho_{max} \sim (a/n)^{1/d+1}$ is the width of the area where one can find backbone or interference scatterer, $r_b \sim a^{1-d/1+d}/n^{2/1+d}$ is the characteristic distance between neighbor backbone impurities. The expression in the first brackets is the characteristic number of scatterers in the hopping act; the expression in the second ones is equal to magnetoresistance in the case of short hops when the real hopping length r_h is substituted by the distance between backbone impurities r_b.

There is a negative magnetoresistance that is linear on H, r_h and $(1 - P_{free})$. And also there is a positive quadratic magnetoresistance due to the alignment of the spins that has strong temperature dependence. On fig. 2 we compare these results with our computations (that include all intermediate impurities and tunneling paths). One can see that analytic results are in good agreement with computations.

4. Discussion

Let us compare our results with experimental data observed in [10, 11]. This experiments show the unusual magnetoresistance dependence on temperature, namely, the suppression of negative magnetoresistance at low temperatures. The spin mechanisms of magnetoresistance was suggested to understand this phenomena in [11], however no detailed theory of this mechanisms was given.

On figure 3 we compared the experimental data with theoretical result (9). One can see that there is a good semiquantitative agreement with experiment.

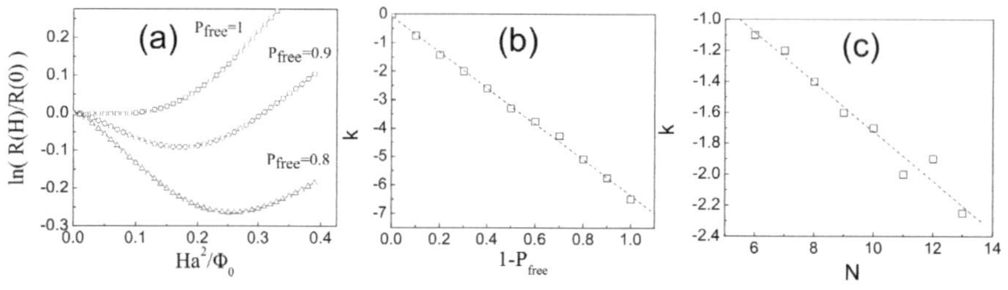

Figure 2. The computation results for the magnetoresistance in the case of long hops. (a) — the magnetoresistance at different P_{free}. (b) — the dependence of linear magnetoresistance coefficient k on $1 - P_{free}$. (c) — the dependance of k on the number of scatterers N which is proportional to the hopping length r_h.

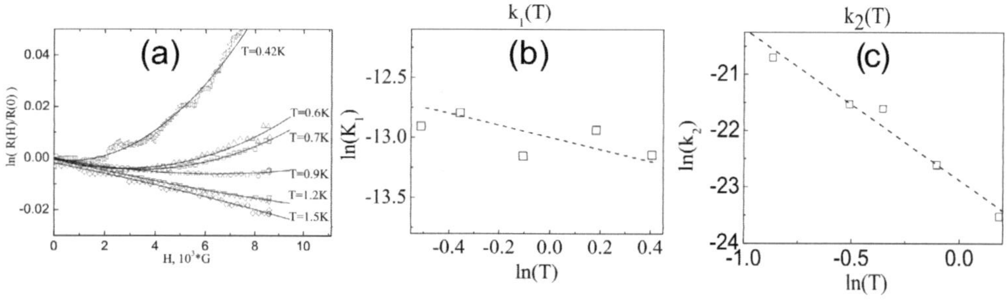

Figure 3. (a) — the experimental data from [10, 11] approximated with parabolic law $-K_1H + K_2H^2$. (b) and (c) — the temperature dependance of K_1 and K_2 correspondingly in the double logarithmic scale. Point are estimated experimental values and lines correspond to our analytic result in the case of long hops (9).

To conclude, we generalize the theory of interference effects on the magnetoresistance in hopping conductivity to include the contribution of scatterers with free electron spins. We considered both the cases of short hops and long hops. For the case of long hops we developed a new approach to the problem of interference magnetoresistance that relies upon the dispersion of scatterer energies E_j. Our theory is in semiquantitative agreement with experimental data on magnetoresistance in $GaAs - AlGaAs$ 2D structures.

We are grateful to A.S. Ioselevich and N.S. Averkiev for many fruitful discussions. A.V.S. acknowledge the support by Dynasty foundation.

[1] B.I. Shklovskii and A.L. Efros, "Electronic Properties of Doped Semiconductors" (Springer, Berlin, 1984).
[2] V.L. Nguyen, B.Z. Spivak, and B.I. Shklovskii, Pisma Zh. Eksp. Teor. Fiz. **41**, 35 (1985); Zh. Eksp. Teor. Fiz. **89**, 1770 (1985) [Sov. Phys. JETP **62**, 1021 (1985)].
[3] Shklovskii B.I., Spivak B.Z. In: Hopping transport in solids, ed. by M.Pollak and B.Shklovskii, Elsevier, 1991, p. 271
[4] H.L.Zhao, B.Z.Spivak, M.P.Gelfand, S.Feng Phys. Rev. B, **44**, 10760 (1991)
[5] W.Schirmacher Phys. Rev B, **41**, 2461 (1990)
[6] M.E. Raikh, G.F. Wessels, Phys. Rev. B, **47**, 15609 (1993)
[7] U.Sivan, O.Entin-Wohlman, Y.Imry, PRL, **60**, 1566 (1988)
[8] Eduardo G. Aponte, Ernesto Medina, PRE, **58**, 4246 (1998)
[9] Yeong-Lieh Lin and Franco Nori, PRL, **76**, 4580 (1996)
[10] N.V.Agrinskaya, V.I.Kozub, D.V. Poloskin, A.V.Chernyaev, D.V.Shamshur JETP Letters **80**, 30 (2004)
[11] N.V. Agrinskaya, V.I. Kozub, A.V. Shumilin, and E. Sobko, Phys. Rev. B, **82** 075201 (2010)
[12] R.N. Bhatt, P.A. Lee, Phys. Rev. Lett., **48**, 344 (1982)
[13] J.R. Eggert, Phys. Rev. B, **26**, 6664 (1984)

Disorder and Interaction Effects in Quantum Wires

L. W. Smith[1], K. J. Thomas[2,*], M. Pepper[3], D. A. Ritchie[1], I. Farrer[1], J. P. Griffiths[1], G. A. C. Jones[1]

[1]Cavendish Laboratory, University of Cambridge, J J Thomson Avenue, Cambridge, CB3 OHE, UK
[2]Dept. of Electronic and Electrical Engineering, Sungkyunkwan University, Suwon 440-746, South Korea
[3]Dept. of Electronic and Electrical Engineering, University College London, London, WC1E 7JE, UK

Abstract

We present conductance measurements of quasi-one-dimensional quantum wires affected by random disorder in a GaAs/AlGaAs heterostructure. In addition to quantised conductance plateaux, we observe structure superimposed on the conductance characteristics when the channel is wide and the density is low. Magnetic field and temperature are varied to characterize the conductance features which depend on the lateral position of the 1D channel formed in a split-gate device. Our results suggest that there is enhanced backscattering in the wide channel limit, which gives rise to quantum interference effects. When the wires are free of disorder and wide, the confinement is weak so that the mutual repulsion of the electrons forces a single row to split into two. The relationship of this topological change to the disorder in the system will be discussed.

Introduction

Quantised conductance is the most important hallmark of a clean one-dimensional (1D) electron system [1, 2]. Deviations from perfect quantization at $G = N\,(2e^2/h)\,T$, where N is the number of transmitting modes and T the transmission coefficient, can occur due to several reasons [3], the most common being scattering from disorder potentials, or imperfections in the devices resulting in $T < 1$. In split-gate devices [4], it is possible to re-establish well quantised plateaux by applying an offset voltage between the two arms of the device which laterally shifts the 1D channel away from a disorder dominated area. This technique was first shown in narrow Silicon channels [5], and Ref. [6] provides an analysis of the change in scattering as the channel is shifted. In practice, by laterally shifting the channel it is possible to check for the existence of localized impurity potentials which may give rise to spurious effects on the sample resistance. Early work on 1D channels showed that the location of a particular scattering centre affected quantised levels in a manner which depended on the parity of the quantised wavefunction [7].

In addition to deviations in the value of conductance plateaux, there can be resonance structures superimposed on the conductance trace. Close to pinch-off there could be periodic conductance oscillations due to tunnelling through a quantum dot which is unintentionally formed by an impurity potential [8]. There can also be interference effects due to electrons being backscattered by potential fluctuations. By shifting the channel sideways, the potential landscape changes, which results in a changing location of structure on the conductance trace. Applying a small perpendicular magnetic field reduces backscattering significantly, and clean characteristics can be reinstated. By increasing temperature, the resonance structures are averaged out. If the resonance is due to imperfections in the device geometry or due to static impurities, moving the channel sideways may not result in clean characteristics and the resonances will be reproducible for each cool down. This cannot be asserted for resonances due to overall potential fluctuations in the crystal, as a second cool down may change the potential landscape and the screening from the local carrier density. Thus by varying the magnetic field, temperature and spatial location of the 1D channel, combined with thermal cycling of the device, it is possible to get an overall knowledge of the origin of the various structures in a conductance measurement.

Journal of Physics: Conference Series **376** (2012) 012018 doi:10.1088/1742-6596/376/1/012018

A difficulty with all these measurements is that there could be electron-electron interaction effects which give rise to deviations in the conductance. In the Luttinger liquid picture, there are disorder-induced deviations in conductance quantization, which has a power-law dependence on temperature and voltage [9, 10]. The most significant interaction effect in 1D ballistic quantum wires is the spin-related structure close to $0.7\times(2e^2/h)$, which has a distinctly different origin to that of a resonance structure. This 0.7 structure does not disappear as the channel is moved sideways, and strengthens as temperature increases (instead of thermally averaging like the disorder-related effects). The 0.7 structure occurs reproducibly in almost all ballistic 1D devices [11]. There are studies on the 0.7 structure relating it to a Kondo effect due to a quasibound state which may form in the 1D wire [12, 13, 14]. However, there are inconsistencies in this argument brought out by later work [15, 16]; for example, Sfigakis et al. [17] have specifically shown that locating a barrier at the exit regions of the 1D channel assists in the creation of a bound state and a resonance structure at $0.5(2e^2/h)$. This structure possesses all the characteristics discussed above for a resonant structure, and shows a Kondo effect with decreasing temperature which appears additive to the 0.7.

In general, it is important to minimize the effects due to disorder in order to investigate pure electron-electron interaction effects. One of the most striking manifestations of the electron interactions is the predicted splitting of a single row of electrons into a double row in order to reduce the mutual repulsion of neighbouring electrons [18, 19]. We have recently shown that gradually reducing the lateral potential which confines the electrons in a 1D wire [20, 21] results in observation of this lateral rearrangement of electrons. In order to obtain an unambiguous identification of this effect we have taken extreme care in checking through several devices before commencing a detailed study on the row formation. This is necessary, because the formation of a double row as a ground state results in a jump from zero conductance to $4e^2/h$, and such behaviour can also be observed if, for example, there is a single potential spike at the centre of the channel, which unintentionally divides the wire into two halves. This was shown by fabricating split-gate devices with an additional 'dot gate' in the centre of the channel, by Smith et al. [22, 23].

In this paper, we investigate the potential landscape of a wide (1 μm) quantum wire, where resonances are superimposed on the conductance characteristics and change their pattern as a function of the channel position. It was found that a small perpendicular magnetic field was sufficient to remove the various structures, and increasing the temperature averages them out. The resonances disappear as the channel is moved to one side and not the other, which indicates that the origin is a disorder potential at one edge of the channel. The pattern was reproduced more or less exactly with different cool downs, indicating that the cause is a static impurity or charged defect. The current study highlights how important it is to fully characterise the devices before conclusions can be drawn about interaction effects.

Sample and Method

Our sample was fabricated from a modulation-doped GaAs/AlGaAs heterostructure. A two-dimensional electron gas (2DEG) was formed 300 nm below the surface of the wafer. The carrier density and mobility of the 2DEG, measured during the second cool down of the device are 1.5×10^{11} cm^{-2} and 1.3×10^{2} V^{-1} s^{-1}, respectively. Split-gate devices with a top gate were patterned using electron-beam lithography. The top gate was 1 μm wide, and was separated from the split gates by a 200 nm layer of cross-linked polymethylmethacrylate (PMMA) [24]. The split gates were 0.4 μm long and 1 μm wide, and a schematic diagram of the device is shown in Fig. 1.

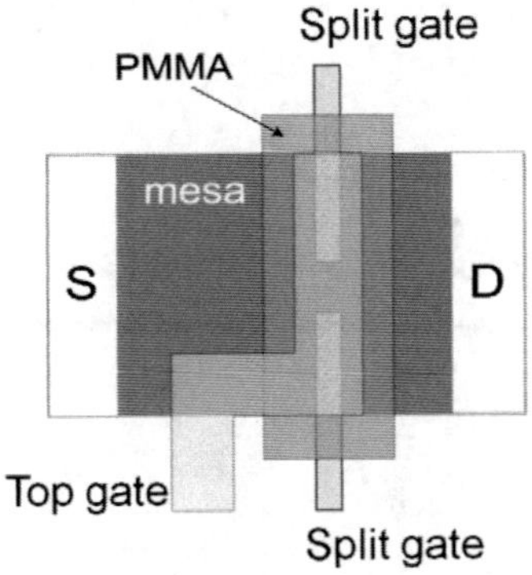

In order to study the reproducibility of conductance features, data are presented from three different cool downs of the same device. For the measurements presented in Fig. 3, the magnetic field was oriented perpendicular to the 2DEG. For another cool down (Fig. 4), the magnetic field was oriented in the plane of the 2DEG, parallel to the 1D channel. Two-terminal measurements were performed using a dilution refrigerator with a sample base temperature of 50 mK, with a 77 Hz excitation voltage of 5 μV. The device is typically operated by applying a fixed negative voltage to the split gates (V_{sg}), and sweeping the voltage applied to the top gate (V_{tg}). Thus, V_{sg} defines the 1D channel, and primarily controls the strength of the confining potential in the transverse direction. With the 1D channel defined, the density in the channel is then varied using V_{tg} [25].

Figure 1 above: Schematic diagram of the device. Three stages of e-beam lithography were required. Firstly, split gates were patterned and metalized. Following this, a 200 nm layer of PMMA was deposited, and cross-linked using the electron beam [24]. Lastly, a top gate was patterned and metalized.

Results and Discussion

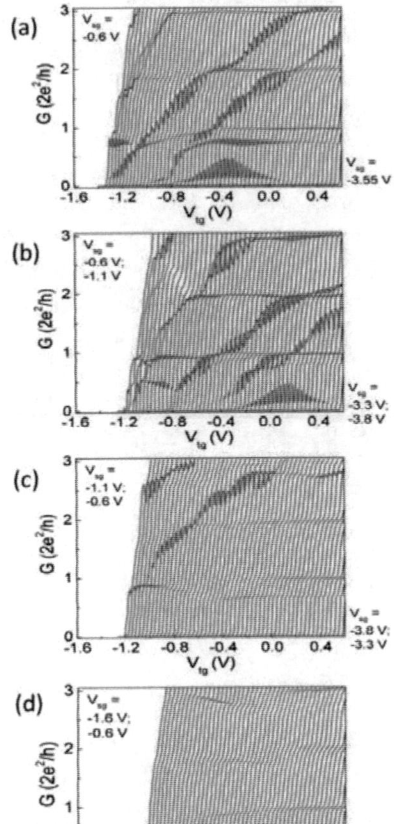

FIGURE 2 left: COOL DOWN 1 - Conductance characteristics as the 1D channel is shifted laterally. The conductance is measured as a function of V_{tg} and V_{sg} is stepped from left-to-right in increments of 50 mV in all the plots. In (a), the offset applied to the two arms of the split gate $\Delta V_{sg} = 0$. In (b), the channel is shifted by applying $\Delta V_{sg} = 0.5$ V. In (c), the 1D channel is shifted in the opposite direction, using $\Delta V_{sg} = -0.5$ V. In (d), ΔV_{sg} is increased to -1 V. The voltage range over which V_{sg} is stepped is shown on each figure. Where an offset is applied, the voltages on both arms of the split gate are given.

In Fig. 2(a), the conductance is measured as a function of V_{tg}, at different fixed V_{sg} with the offset applied to the two arms of the split-gate $\Delta V_{sg} = 0$. Between traces, V_{sg} is stepped by 50 mV, from -0.6 V on the left to -3.55 V on the right. As V_{sg} is made more negative, the electrostatic potential confining the 1D channel becomes steeper and the channel becomes narrower. Therefore, following the conductance traces from left-to-right is equivalent to moving from a weakly-confined, wider, regime to a more strongly-confined regime. The first three quantized plateaus can be clearly observed to the right of the figure in the strong-confinement regime, (more negative V_{sg}). In addition to the plateaux, several trails of resonance structures can be seen, emanating from the left-hand side of the figure (weak-confinement regime), which move diagonally across to higher conductance. The resonances in general are weaker at the plateau region and stronger midway between plateaux, and appear to "form" plateaux themselves coincident with the original plateaux.

In (b), an offset $\Delta V_{sg} = 0.5$ V is applied, which laterally shifts the 1D channel in one direction. V_{sg} is stepped from -0.6, -1.1 V on the left-hand side of the plot, to -3.3, -3.8 V on the right-hand side (voltages on both arms of the split gate are given). In this case, all features are qualitatively present as in (a), but are shifted to the right-hand side of the figure. In (c), the channel is shifted in the opposite direction where ΔV_{sg} = -0.5 V; V_{sg} is stepped from -1.1, -0.6 V on the left to -3.8, -3.3 V on the right. The channel is shifted further in the same direction in (d), where ΔV_{sg} = -1 V is applied, and V_{sg} is stepped from -1.6, -0.6 V on the left of the plot to -4, -3 V on the right.

In Figs. 2(c) and (d), it is clear that the strength of the resonances are diminishing, indicating that the channel is moving away from a disorder dominated landscape. A large offset of -1 V has produced cleaner conductance characteristics in (d); however, this has also weakened the conductance plateaux, possibly due to the decreased 1D subband spacings.

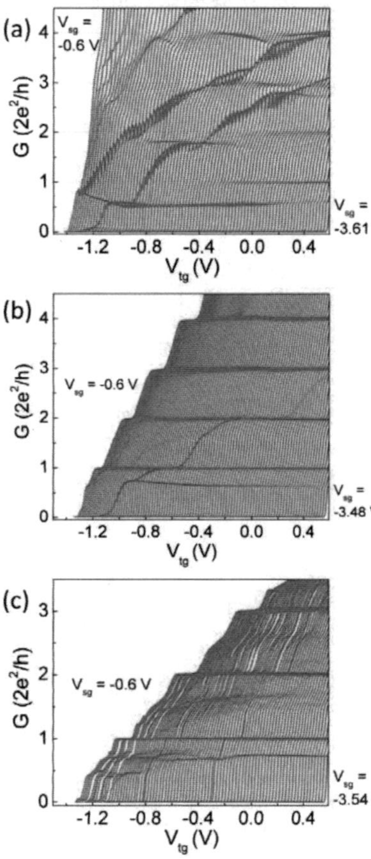

FIGURE 3 left: COOL DOWN 2 - Conductance characteristics as a function of V_{tg} and V_{sg} at a fixed perpendicular magnetic field, ΔV_{sg} = 0. (a) $B_\perp$ = 0; V_{sg} is stepped from -0.6 V on the left to -3.61 V on the right, in intervals of 35 mV. (b) $B_\perp$ = 0.4 T; V_{sg} is stepped from -0.6 V on the left to -3.48 V on the right, in intervals of 30 mV, and (c) $B_\perp$ = 0.8 T; V_{sg} is stepped from -0.6 V on the left to -3.54 V on the right, in intervals of 35 mV.

Figures 3(a) to (c) show conductance data from a second cool down of the same device, where ΔV_{sg} = 0. In (a), V_{sg} is stepped from -0.6 V on the left to -3.61 V on the right. In this cool down, the trails of resonance features are more or less reproduced. In (b), a perpendicular magnetic field ($B_\perp$) of 0.4 T is applied, and V_{sg} is stepped from -0.6 V to -3.48 V (left-to-right). The resonance features are suppressed quite remarkably, and the additional confinement from the magnetic field has enhanced the quality of the plateaux, particularly in the weak-confinement regime (left-hand side). It is noteworthy that the one trail of resonances which remains in (b) forms a plateau at $0.7(2e^2/h)$.

Scattering by an impurity potential is reduced by the edge states which are established in a perpendicular field. At $B_\perp$ = 0.8 T [Fig. 3(c)] the plateaux show features related to the quantum Hall effect, and almost all of the resonances have disappeared. Spin splitting is weakly discernible on several traces on the left-hand side, where the confinement is weaker and the 1D subband spacings are smaller. The 0.7 structure prevails throughout the different confinement regimes, in addition a plateau appears at e^2/h in Fig. 3(c), in the weak-confinement regime.

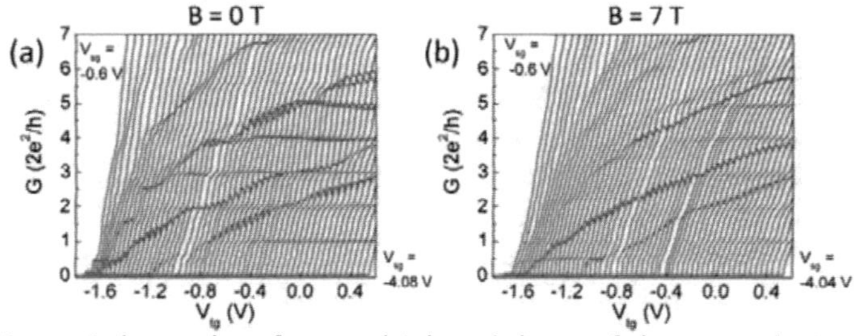

FIGURE 4: COOL DOWN 3 - Conductance characteristics from a third cool down, in which the magnetic field is applied in the plane of the 2DEG, parallel to the 1D channel. Offset ΔV_{sg} = 0. In (a), B = 0 T, and V_{sg} is stepped from -0.6 V on the left to -4.08 V on the right, in 60 mV intervals. In (b), B = 7 T, and V_{sg} is stepped from -0.6 V on the left to -4.04 V on the right, also in 60 mV intervals.

Figure 4 shows data from a third cool down of the same device. At B = 0, the resonance trails are present with a slightly different pattern to Figs. 2(a) and 3(a). In addition to the trails, the disappearance of the first plateau (at $2e^2/h$), and a direct jump of $4e^2/h$ can be observed at V_{tg} ≈-0.9 V, although this is less clear than when the conductance characteristics are clean. The jump to $4e^2/h$ marks the formation of a double row of electrons, discussed in our previous work [21]. However, without a full knowledge of the pattern of resonances that are superimposed on the conductance traces, it is difficult to attribute the disappearance of the first quantised plateau to double-row formation. At an applied in-plane magnetic field B = 7 T [Fig. 4(b)], the resonance trails are present though slightly suppressed, perhaps related to reduced transmission due to the shrinkage of the wavefunction from a magnetic field parallel to the channel axis. Spin splitting is hardly discernible, except in a small range of gate voltages from V_{tg} = -0.9 V to -1.5 V where there is structure at e^2/h. This cannot be a trail of resonance as such a trail was absent in zero field [Fig. 4(a)]. If it were a spin-split plateau, this is an indication of an absence of row formation in an in-plane magnetic field, due to disorder. All higher resonant trails are mostly unaffected, unlike the perpendicular-field case [shown in Figs. 3(b) and 3(c)]. In general it seems that the existence of the disorder is obscuring the observation of the double-row ground state.

Figure 5 shows a conductance measurement at T = 1 K and ΔV_{sg} = 0, from the first cool down. As expected, there is an overall thermal averaging of the conductance features at this elevated temperature, and the resonant features are suppressed. A weak first plateau is visible at higher confinement strengths, along with the 0.7 structure. It is surprising that the resonance structures are stronger at higher conductance, and that those which disappear first are towards the strong-confinement regime. Generally it is expected that the structures at stronger confinement prevail with increasing temperature. The contrary nature of the data suggests that the features discussed in this paper are not due to resonant transmission of quasi-bound states in the channel, but rather are interference phenomena due to backscattering of electrons in the wide channel limit. As the channels narrows, backscattering is suppressed due to the rigid one-dimensionality. In a perpendicular magnetic field, the opposite is the case [Figs. 3(b) and 3(c)]: the resonance trails on the left-hand-side (weak confinement regime) disappear first as the edge states are further apart than in the strong-confinement regime, where there can be scattering between edge states.

Data from this same device are presented in Ref. [21], for which an offset of -2 V was applied to the split gate, further shifting the 1D channel. This lateral shift resulted in clean conductance characteristics across the whole confinement regime, and we conducted a detailed study on the formation of a coupled row of electrons which formed in the weak-confinement regime. Our results were confirmed by additional measurements using a similar, but disorder-free device, with no offset applied between the arms of the device [26].

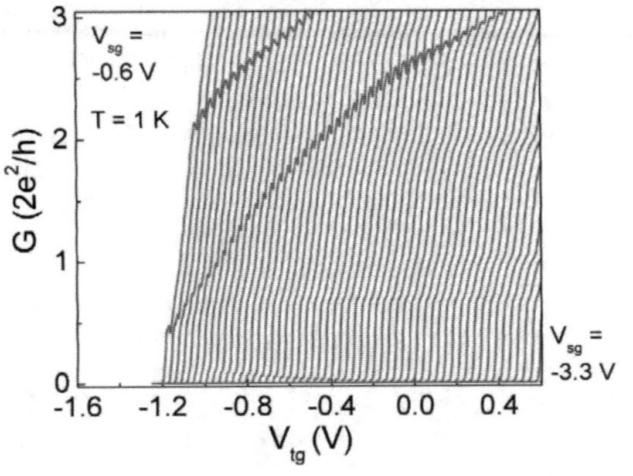

FIGURE 5: COOL DOWN 1 – These are the only results above 50mK. Conductance characteristics at T = 1 K, for ΔV_{sg} = 0. V_{tg} is swept for fixed V_{sg}, and V_{sg} is stepped between traces from -0.6 V on the left to -3.3 V on the right. These data were obtained 8 days after the data presented in Fig. 2, therefore there is a slight drift in pinch-off characteristics, and the resonances do not occur at exactly the same gate voltages.

Conclusion

In conclusion, we have shown that a potential fluctuation within or near the path of a 1D channel gives rise to resonant reflections, manifested by a trail of resonances in the conductance. Moving the channel sideways by applying an offset voltage to arms of the the split gate ensures the channel is shifted further from the impurity potential. This progressively removes the disorder and produces clean conductance characteristics. A clear observation of interaction effects leading to the formation of a double-row ground state is reduced by the presence of the disorder.

Acknowledgements

This work was supported by the Engineering and Physical Sciences Research Council (UK). K.J.T. acknowledges support from WCU program through the NRF of Korea funded by the Ministry of Education, Science and Technology (R32-2008-000-10204-0). We thank F. Sfigakis for helpful discussions.

References:

*Email address: kalarikad@skku.edu

[1] D. A. Wharam, T.J. Thornton, R. Newbury, M. Pepper, H. Ahmed, J. E. F. Frost, D. G. Hasko, D. C. Peacock, D. A. Ritchie and G. A. C. Jones, *One dimensional transport and the quantisation of the ballistic resistance*, *J. Phys. C* **21**, L209, (1988).
[2] B. J. van Wees, H. van Houten, C. W. J. Beenakker, J. G. Williamson, L. P. Kouwenhoven, D. van der Marel, and C. T. Foxon, *Quantized conductance of point contacts in a two-dimensional electron gas*, Phys. Rev. Lett. **60**, 848 (1988)
[3] G. Timp, R. Behringer, S. Sampere, J. E. Cunningham, and R. E. Howard, *When isn't the conductance of an electron waveguide quantized?* in *Proc. Int. Symp. on Nanostructure Physics and Fabrication,* W.P. Kirk and M. Reed, Eds. New York: Academic, (1989), 331-345.
[4] T. J. Thornton, M. Pepper, H. Ahmed, D. Andrews, and G. J. Davies, *One-Dimensional Conduction in the 2D Electron Gas of a GaAs-AlGaAs Heterojunction*, Phys. Rev. Lett. **56**, 1198 (1986).
[5] R. J. Stroh and M.Pepper, *Conductance fluctuations from the local alteration of a hopping path*, J. Phys.: Condens. Matt. **1**, 8481 (1989).
[6] L. I. Glazman and I. A. Larkin, *Lateral position control of an electron channel in a split-gate device*, Semicond. Sci. Technol. **6**, 32 (1991).
[7] J. G. Williamson, C. E. Timmering, C. J. P. M. Harmans, J. J. Harris, and C. T. Foxon, *Quantum point contact as a local probe of the electrostatic potential contours*, Phys. Rev. B **42**, 7675 (1990).
[8] J. T. Nicholls, J. E. F. Frost, M. Pepper, D. A. Ritchie, M. P. Grimshaw, and G. A. C. Jones, *Charging effects and the excitation spectrum of a quantum dot formed by an impurity potential*, Phys. Rev. B **48**, 8866 (1993).
[9] M. Ogata and H. Fukuyama, *Collapse of Quantized Conductance in a Dirty Tomonaga-Luttinger Liquid*, Phys. Rev. Lett. **73**, 468 (1994).
]10] S. Tarucha, T. Honda, and T. Saku, *Reduction of quantized conductance at low temperatures observed in 2 to 10 μm-long quantum wires*, Solid State Commun. **94**, 413 (1995).
[11] K. J. Thomas, J. T. Nicholls, M. Y. Simmons, M. Pepper, D. R. Mace, and D. A. Ritchie, *Possible Spin Polarization in a One-Dimensional Electron Gas*, Phys. Rev. Lett. **77**, 135 (1996).
[12] S. M. Cronenwett, H. J. Lynch, D. Goldhaber-Gordon, L. P. Kouwenhoven, C. M. Marcus, K. Hirose, N. S. Wingreen, and V. Umansky, *Low-Temperature Fate of the 0.7 Structure in a Point Contact: A Kondo-like Correlated State in an Open System,* Phys. Rev. Lett. **88**, 226805 (2002).
[13] Y. Meir, K. Hirose, and N. S. Wingreen, *Kondo Model for the "0.7 Anomaly" in Transport through a Quantum Point Contact*, Phys. Rev. Lett. **89**, 196802 (2002).
[14] T. Rejec and Y. Meir, *Magnetic impurity formation in quantum point contacts*, Nature **442**, 900 (2006).
15] A. C. Graham, K. J. Thomas, M. Pepper, N. R. Cooper, M. Y. Simmons, and D. A. Ritchie, *Interaction Effects at Crossings of Spin-Polarized One-Dimensional Subbands*, Phys. Rev. Lett. **91**, 136404 (2003).
[16] A. C. Graham, D. L. Sawkey, M. Pepper, M. Y. Simmons, and D. A. Ritchie, *Energy-level pinning and the 0.7 spin state in one dimension: GaAs quantum wires studied using finite-bias spectroscopy*, Phys. Rev. B **75**, 035331 (2007).
[17] F. Sfigakis, C. J. B. Ford, M. Pepper, M. Kataoka, D. A. Ritchie, and M. Y. Simmons, *Kondo Effect from a Tunable Bound State within a Quantum Wire*, Phys. Rev. Lett. **100**, 026807 (2008).
[18] A. D. Klironomos, J. S. Meyer, and K. A. Matveev, *Spontaneous spin polarization in quantum wires*, Europhys Lett. **74**, 679 (2006).
[19] J. S. Meyer, K. A. Matveev, and A. I. Larkin, *Transition from a One-Dimensional to a Quasi-One-Dimensional State in Interacting Quantum Wires*, Phys. Rev. Lett. **98**, 126404 (2007).
[20] W. K. Hew, K. J. Thomas, M. Pepper, I. Farrer, D. Anderson, G. A. C. Jones, and D.A.Ritchie, *Incipient Formation of an Electron Lattice in a Weakly Confined Quantum Wire*, Phys. Rev. Lett. **102**, 056804 (2009).

[21] L. W. Smith, W. K. Hew, K. J. Thomas, M. Pepper, I. Farrer, D. Anderson, G. A. C. Jones, and D. A. Ritchie, *Row coupling in an interacting quasi-one-dimensional quantum wire investigated using transport measurements*, Phys. Rev. B **80**, 041306 (2009).
[22] C. G. Smith, M. Pepper, R. Newbury, J. Ahmed, D. G. Hasko, D. C. Peacock, J. E. F. Frost, D. A. Ritchie, G. A. C. Jones, and G. Hill, *One-dimensional quantised ballistic resistors in parallel configuration*, J. Phys.: Condens. Matter **1**, 6763 (1989).
[23] P. Simpson, D. Mace, C. Ford, I. Zailer, M. Pepper, D. Ritchie, J. Frost, M. Grimshaw, and G. A. C. Jones, *Aharonov-Bohm effect and one-dimensional ballistic transport through two independent parallel channels*, Appl. Phys. Lett. **63**, 3191 (1993).
[24] I. Zailer, J. E. F. Frost, V. Chabasseur-Molyneux, C. J. B. Ford, and M. Pepper, *Crosslinked PMMA as a high-resolution negative resist for electron beam lithography and applications for physics of low-dimensional structures*, Semicond. Sci. Technol. **11**, 1235 (1996).
[25] W. Hew, K. Thomas, I. Farrer, D. Anderson, D. Ritchie, and M. Pepper, *Tuning the confinement strength in a split-gate quantum wire*, Physica E **40**, 1645 (2008).
[26] L. W. Smith, W. K. Hew, K. J. Thomas, M. Pepper, I. Farrer, D. Anderson, G. A. C. Jones, and D. A. Ritchie, *Coupled double-row formation in a quasi-1D wire*, Physica E **42**, 1114 (2010).

Journal of Physics: Conference Series **376** (2012) 012019 doi:10.1088/1742-6596/376/1/012019

Fano-Kondo oscillations of the conductance and thermopower in a mesoscopic transistor

A. Aldea[1,2], M. Ţolea[1], and I.V. Dinu[1]

[1] National Institute of Materials Physics, 77125 Bucharest-Magurele, Romania
2 Institute of Theoretical Physics, Cologne University, 50937 Cologne, Germany.

E-mail: aldea@infim.ro

Abstract. We study the conductance G and Seebeck coefficient S of a side-coupled double quantum dot system. The transport properties are the result of the competition between the Fano interference and Kondo correlations, being also controlled by the Coulomb blockade of the multilevel side-dot. The external parameters are the gate potential V_g applied on the side-dot and the temperature. When V_g is varied continuously the blockade is switched on and off periodically, resulting in oscillations of the transport coefficients. The profile of the oscillations and the temperature dependence are studied. An extended Anderson model and the Keldysh transport formalism are used.

We investigate the transport properties of a double quantum dot system arranged in a transistor-like setup as it is shown in Fig.1. A small interacting dot connected to external leads is pre-prepared in the Kondo state using the gate V_{g0}, while a large side-dot containing many energy levels plays the role of the transistor basis. The functional parameter of the device is the gate potential V_g applied on the side-dot. When this potential is varied continuously the Coulomb blockade of the side dot is switched on/off periodically giving rise to oscillations of the source-drain current with a period equal to the charging energy. The reason for oscillations is the periodic change of the interference conditions in the double dot device.

Different aspects of the side coupled double dots have already been studied. We mention the two-stage problem [1], the two channel problem [2] and the Fano-Kondo effect [3,4] which are specific to this geometry. Non-equilibrium properties have also been discussed; the possibility of the peak-dip crossover of the differential conductance around the zero bias and questions concerning the scaling properties of the non-equilibrium conductance have been addressed recently [5].

The charge current through the device is the result of the Fano interference between the resonant channel passing through the lateral dot and the continuous channel through the small dot connected directly to the leads. The interference process is strongly affected by the regime (Kondo or non-Kondo) of the small dot, and vice-versa the Kondo effect may be suppressed by destructive Fano interference [4,5]. As a consequence of the competition between the two processes the oscillations of the current should show a Fano-Kondo profile with an asymmetry factor which depends on the weight of the two channels acting in the interference process.

The processes discussed above affect not only the linear conductance of the device but also the thermoelectric coefficient. The energy current should include in principle electron, phonon and interaction contributions. The interactions being essential in mesosystems, their presence induces specific features of the thermopower as the violation of the Wiedemann-Franz law and the possibility of a high figure of merit ZT [6]. We mention from the very beginning that our study considers only the coherent electronic contribution to thermopower. Inelastic processes have been studied in [7].

For single dots, the calculations by Costi and Zlatic [8] indicate the possibility of a sign changing of the Seebeck coefficient with increasing temperature, in agreement with the experimental facts [9]. For multiple quantum dots the problem has not yet been studied and this paper is a tentative to identify the specific changes in the themopower of mesoscopic systems resulting from the combined action of correlation and interference processes.

The phenomena occurring in the mesoscopic transistor will be described in terms of an extended Anderson model, using the Keldysh transport formalism and equation of motion technique for the calculation of the Green functions.

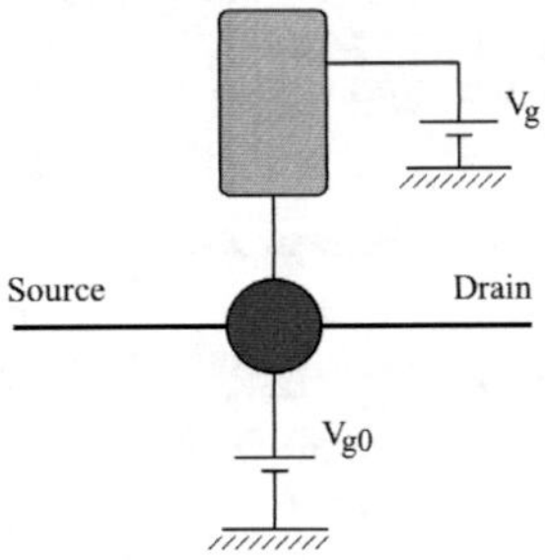

Figure 1. The sketch of the double dot system: a small dot in Kondo regime is connected to external leads, and to a non-interacting multilevel lateral dot. The gate potential V_g applied on the lateral dot is the functional parameter of this transistor-like device.

The set-up shown in Fig.1 can be described by the following extended Anderson model:

$$H = \sum_{k,\sigma,\alpha} (\epsilon_k - \mu_\alpha) c^\dagger_{k\sigma,\alpha} c_{k\sigma,\alpha} + \sum_\sigma E_d d^\dagger_\sigma d_\sigma + U_H n_{d\uparrow} n_{d\downarrow} + \sum_{i\sigma} E_i c^\dagger_{i\sigma} c_{i\sigma} + \frac{e^2}{2C} N^2$$
$$+ \sum_{k,\sigma,\alpha} t_{kd} (c^\dagger_{k\sigma,\alpha} d_\sigma + h.c.) + \sum_i \tau_{id} (c^\dagger_{i\sigma} d_\sigma + h.c.), \tag{1}$$

where the terms describe in turn the leads, the Kondo dot with the Hubbard interaction, the lateral dot and Coulomb blockade interaction, and finally the two couplings. The parameters t and τ describe the lead-dot and interdot tunnel coupling, respectively.

It is important to note that we choose a specific working regime of the system defined by

$$T < T_k < \delta E < U_C, \tag{2}$$

situation which can be met experimentally, but does not describe the 2-channel the 2-stage cases (T = temperature, T_K = Kondo temperature; δE and $U_C = \frac{e^2}{2C}$ are the interlevel spacing and charging energy of the big dot).

For the considered geometry, the quantity needed to calculate the transport properties is the Green function $G_{dd}(\omega)$ of the small dot (indexed by d). For the calculation of $G_{dd}(\omega)$ we use the equation of motion technique in the version of Ref.10 applicable to complex mesostructures. The input quantity is the non-interacting self-energy Σ_0, which in our case includes two terms corresponding to the two couplings in eq.(1)

$$\Sigma_0(\omega) = \sum_{k,\alpha} \frac{|t_{kd}|^2}{\omega^+ - \epsilon_k + \mu_\alpha} + \sum_i \frac{|\tau_{id}|^2}{\omega^+ - E_i} = -i\Gamma + \sum_i \frac{|\tau_{id}|^2}{\omega^+ - E_i}\ . \tag{3}$$

One notices that the two contributions are additive and both of them acts on the shape of the Kondo peak of the density of states of the small dot. The second term is that one which gives rise to the *Kondo in a box* effect, which was studied in [11] for the case of vanishing dot-leads coupling.

The charge current flowing through the device is given in the Keldysh formalism by the well-known formula:

$$I(V) = \frac{2e}{\hbar} \int d\omega [f_L(\omega) - f_R(\omega)]\ \Gamma(\omega)[-\frac{1}{\pi}\ ImG^r_{dd}(\omega)]\ . \tag{4}$$

In the linear approximation with respect to $\delta V = V_L - V_R$ and $\delta T = T_L - T_R$, the charge current can be written as

$$I = L_{11}\delta V + L_{12}\delta T, \tag{5}$$

where, assuming that Γ is independent of ω, the two linear kinetic coefficients L_{11} (the conductance) and L_{12} (the thermoelectric coefficient) are given by

$$L_{11} = \frac{e^2\Gamma}{h} \int d\omega(-\frac{df}{d\omega})ImG_{dd}(\omega), \quad L_{12} = \frac{e\Gamma}{hT} \int d\omega\omega(-\frac{df}{d\omega})ImG_{dd}(\omega). \tag{6}$$

Experimentally, besides the conductance $G = L_{11}$, one measures the Seebeck coefficient $S = \delta V/\delta T = -L_{12}/L_{11}$ which is obtained at vanishing current $I = 0$.

One should notice that L_{11} is determined mainly by the magnitude of the Kondo resonance, while the thermoelectric coefficient L_{12} depends on the shape asymmetry of the resonance around the Fermi energy at the given temperature. The Anderson model and the numerical renormalization group (NRG) method have been used in Ref.8 for the study of thermopower in single quantum dots. Although the equation of motion technique (EOM) which we use in this paper is not reliable in the limit of very low temperatures, the results obtained by this metod for single dots are qualitatively similar to those obtained in the aforementioned references. This fact is encouraging for using the EOM method also in the case of double dots.

In the side-coupled double dot the spectral function at the site 'd' depends not only on the temperature and the parameters describing the Kondo dot but also on the state of the side-dot which is controlled by the gate potential V_g. Compared to the spectral function of the single dot, the spectral function is strongly modified in this case especially when the energy levels of the side dot are in the vicinity of the d-level of the Kondo dot [5]. Such a situation is shown in Fig.2.

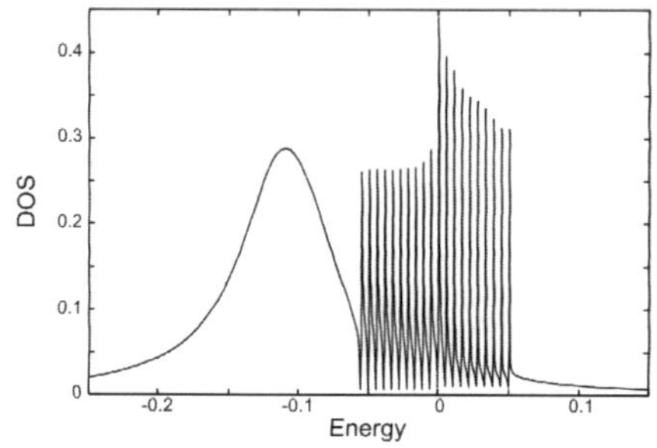

Figure 2. The local density of states at the site d, $DOS = -\frac{1}{\pi}ImG_{dd}$ of the double dot in Kondo regime for the gate potential $Vg = -0.05$ ($E_d = -0.1, \Gamma = 0.02, \tau = 0.005, T = T_k/10$; the energy unit is the half-width of the leads band) [5].

As we have mentioned, by switching on/off the Coulomb blockade of the big dot, the interference conditions are modified periodically in the meso-transistor device. Accordingly, one expects oscillations of the 'source-drain' conductance as function of the gate potential V_g on the big dot; the device should behave like a mesoscopic transistor oscillator. The result of the numerical calculation of the conductance is shown in Fig.3, where the oscillations represent a sequence of Fano profiles which repeat itself with the periodicity $\Delta = \delta E + U_C$. The eventual overlapping of consecutive Fano resonances depends on the ratio τ/Δ; if the ratio is small the resonances are independent and the *background* of the Fano line is quite visible (this is the case in Fig.3). This property is quite understandable since a small interdot coupling τ means a narrow resonance, and a large Δ means that the resonances are much apart. The Fano asymmetry factor q, which in the non-interacting case $U_H = 0$ is simply $q = E_D/\Gamma$, depends now also on temperature due to the presence of the Kondo channel (which, obviously, is temperature dependent). We have found that when the Kondo channel conducts well, i.e at low temperatures, the asymmetry factor q is small.

One may notice in Fig.3 that the temperature dependence of the Kondo-type, meaning the increase of the conductance with decreasing temperature, occurs in the middle of the period, while near the Fano *zero* the Kondo effect is suppressed. This result is in agreement with the experimental data of Sasaki et al [4]. The Fano *zeros* occur at $V_g = n\Delta$ (n=integer), and, by calculating the electron occupancy of the big dot, one can check that these values of V_g correspond to the degeneracy points in the blockade process (where the occupancy of the big dot changes abruptly by one electron).

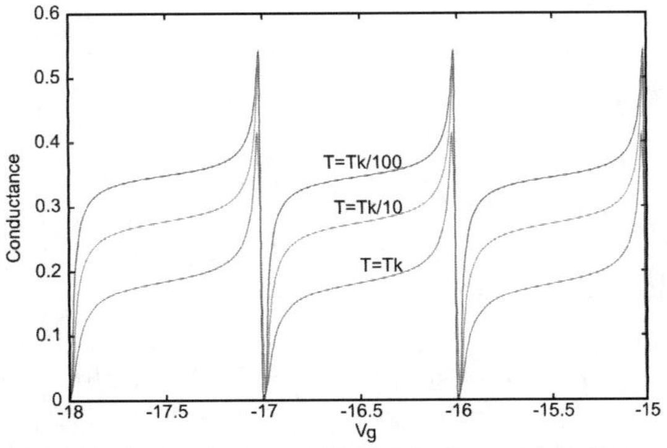

Figure 3. Three periods of the oscillations in the Kondo regime exhibiting Fano profile; the curves correspond to different temperatures: $T/TK = 1/100, 1/10, 1$. The number of levels in the big dot is N=20; the other parameters are: $E_d = -0.2, \tau = 0.005, \Gamma = 0.04$ (the energy unit is the half-width of the lead band).

One may ask the question how the thermoelectric properties are affected by the interference process specific to the side-coupled double dot. The numerical calculation -the only one possible for interacting systems- shows that in the range where the Kondo effect is strong (i.e. in the middle of the oscillation period) the Seebeck coefficient is small indicating that the asymmetry of the density of states is also small. However strong oscillations accompanied by the change of sign occur around the Fano *zero*, i.e. in the range where the Kondo effect is suppressed (see Fig 4a). The temperature dependence of the Seebeck coefficient shown in Fig 4b shows also the possibility of sign changing, which occurs around the Kondo temperature suggesting significant modifications of the shape of the density of states with vanishing correlation effect. It is interesting to notice that the thermoelectric coefficients L_{12} and S do not exhibit a Fano profile as the conductance G does.

In conclusion, the transport coefficients of the side-coupled double dot show specific properties as function on the gate potential V_g, temperature T, and interdot tunnel coupling τ as an outcome of the competition between Kondo correlation and Fano interference. More than that, the periodic blockade of the side-dot occurring at the variation of the gate gives rise to oscillations of the linear response coefficients. The oscillations have a period equal to the charging energy and shows a Fano profile described by an asymmetry factor which increases with increasing temperature. The Fano *zeros* occur at values of the gate corresponding to the degeneracy points of the Coulomb blockade process (i.e. when the side-dot is unblocked); it turns out that around these points the Kondo correlations are suppressed, similar to the experimental finding. The Seebeck coefficient depends on the asymmetry of the local density of states around E_F, and proves to be suppressed by the Kondo correlations and to get large values around the Fano *zeros*. Change of sign of the thermopower occurs at values of the gate potential corresponding to the degeneracy points and at temperatures higher than T_K.

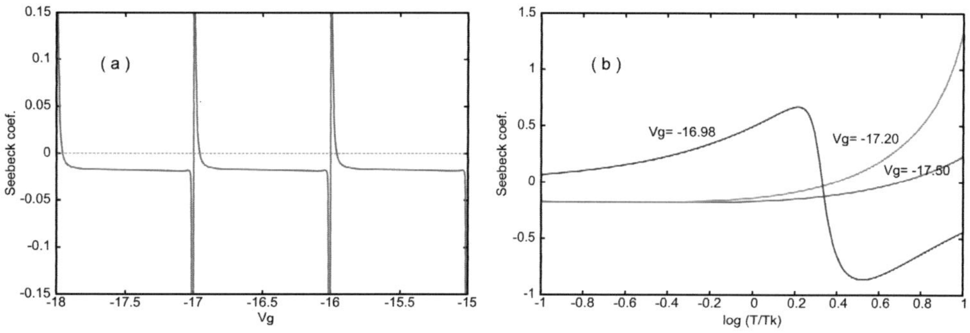

Figure 4. (a) The Seebeck coefficient as function of the gate potential V_g for $T = T_K/10$. (b) Temperature dependence of the Seebeck coefficient for various V_g, chosen in the middle of the oscillation period ($V_g = -17.5, -17.2$; the values are amplified $10\times$ on the y-axis) and near the Fano *zero* ($V_g = -16.98$); V_g is measured in units Δ.

Acknowledgments We acknowledge support from PNCDI2-Research Programme (grant no 515/2009), and Sonderforschungsbereich 608 at the Institute of Theoretical Physics, University of Cologne. One of the authors (AA) is much indebted to Theo Costi for very useful discussions.

References

[1] Cornaglia P S, and Grempel D R 2005 *Phys.Rev.B* **71** 075305 ; C.H. Chung C H, G. Zarand G, and Wölfle P, 2008 *Phys.Rev.B* **77** 035120
[2] Potok R M, Rau I G, Shtrikman H, Oreg Y, and Goldhaber-Gordon D 2007 *Nature* **446** 167
[3] Hofstetter W, König J, and Schoeller H 2001 *Phys.Rev.Lett.* **87** 156803 ; Wu B H, Cao J C, and Ahn K H 2005 *Phys.Rev.B* **72** 165313
[4] Sasaki S ,Tamura H, Akazaki T, and Fujisawa T 2009 *Phys.Rev.Lett.* **103** 266806
[5] Aldea A, Tolea M, and Dinu I V 2011 *Phys.Rev.B* **83** 245317
[6] Liu J, Sun Q, Xie X C 2010 *Phys.Rev.B* **81** 245323
[7] Entin-Wohlman O, Imry Y, Aharony A, 2010 *Phys.Rev.B* **82** 115314
[8] Costi T A, Zlatic V 2010 *Phys.Rev.B* **81** 235127; Costi T A and Hewson A C 1993 *J. Phys.: Condens. Matter* **5** L361
[9] Scheibner R, Buhmann H, Reuter D, Kiselev M N, and Molenkamp L W 2005 *Phys.Rev.Lett.* **95** 176602
[10] Entin-Wohlman O, Aharony A, and Y. Meir Y 2005 *Phys.Rev. B* **71** 035333
[11] Thimm W B, Kroha J, and von Delft J 1999 *Phys.Rev.Lett.* **82** 2143 ; Simon P, Salomez J, and Feinberg D 2006 *Phys. Rev. B* **73** 205325

Distribution of the Entanglement Entropy in Disordered One-Dimensional Fermions

Richard Berkovits

Department of Physics, Bar-Ilan University, Ramat-Gan 52900, Israel

E-mail: berkov@mail.biu.ac.il

Abstract. The entanglement entropy is related to quantum phase transitions. It behaves differently when a quantum system is in a phase with a finite correlation length (such as an insulator phase), compared to a system in a critical phase in which the correlation length is infinite such as a metallic phase. The entanglement entropy of disordered systems does not depend only on the length of the relevant segment and the size of the system but also on the specific realization of disorder. Thus, the entanglement entropy of a disordered system exhibits mesoscopic fluctuations. Here we shall show numerical results that exhibit a crossover in the distribution of the entanglement entropy from a normal distribution at the weak localization regime to a log-normal distribution at the strong localization regime.

1. Introduction

Entanglement entropy (EE) has garnered much recent interest [1]. For condensed matter physics, perhaps the most interesting property related to the EE is its behavior in the vicinity of a quantum phase transition (QPT) [1, 2, 3, 4]. As usual the EE for a system described by a pure state $|\Psi\rangle$ measures the entanglement between two regions A and B composing the system. Specifically, The EE S_A of a region A is defined by using the eigenvalues λ_i of its reduced density operator $\hat{\rho}_A = \mathrm{Tr}_B|\Psi\rangle\langle\Psi|$ (where Tr_B is a trace over all degrees of freedom belonging to region B), as

$$S_A = -\Sigma_i \lambda_i \ln(\lambda_i). \tag{1}$$

For an interacting one-dimensional electronic system S_A will grow logarithmic with the size of region A, L_A, if the system is in the metallic regime (i.e., has no disorder), while for insulators (disordered systems) it will saturate once the size of region A will exceed the localization length ξ [5]. Other aspects of the crossover in the EE of one-dimensional electronic systems can be utilized to identify and study various QPT, such as the phase transition predicted for the population switching in quantum dots coupled to Luttinger liquid leads [6].

Using the concept of EE in the framework of disordered systems raises, as usual in this context, the question of fluctuations and averaging. As with the conductance through a disordered system, one would expect also the EE to show mesoscopic fluctuation. This immediately leads to the question what is the form of the distribution of the EE, and does it depend of the regime of the system, i.e., whether the system length L or the segment length L_A is much smaller than localization length ξ or vise-versa.

In this paper we will show some numerical evidence that the EE distribution crosses over from a normal distribution in the weak localization regime ($\xi \gg L, L_A$) to a log-normal distribution in the strongly localized regime ($\xi \ll L, L_A$). This crossover is reminiscence of the behavior of the distribution of the conductance in one- and higher dimensional disordered systems [12].

2. Model

The Hamiltonian of the one-dimensional spinless electronic system of size L with repulsive nearest neighbor interactions and with on-site disorder is given by the Anderson model:

$$H = \sum_{j=1}^{L} \epsilon_j \hat{c}_j^\dagger \hat{c}_j - t \sum_{j=1}^{L-1} (\hat{c}_j^\dagger \hat{c}_{j+1} + h.c.) + U \sum_{j=1}^{L-1} (\hat{c}_j^\dagger \hat{c}_j - \frac{1}{2})(\hat{c}_{j+1}^\dagger \hat{c}_{j+1} - \frac{1}{2}), \quad (2)$$

for which $\hat{c}_j^\dagger$ ($\hat{c}_j$) is the creation (annihilation) operator of a spinless electron at site j. The random on-site energies are given by ϵ_j drawn from a uniform distribution $[-W/2, W/2]$, U is the nearest-neighbour interaction strength where a positive background is included, and the hopping matrix element t is taken as unity. For the non-interacting case the localization length is known to follow $\xi(W, U = 0) \approx 105/W^2$. When interactions are considered $\xi(U) = (\xi(U = 0))^{1/(3-2g(U))}$ [7] where the Luttinger liquid parameter $g(U) = \pi/[2\cos^{-1}(-U/2)]$ [8]. Since for repulsive interactions $g < 1$ decreases as a function of the interaction strength, the localization length decreases as a function of the interaction strength [9].

DMRG (Density matrix renormalization group) [10, 11] will be used in order to numerically calculate the ground state EE of the Hamiltonian in Eq. (2) for different realizations of disorder, (600 realizations for the shortest samples, 400 for the longer ones), different length $L = 300, 500, 700$ and different interaction strength $U = 0, 0.6$ (corresponding to $g = 1, 0.836$).

3. Results

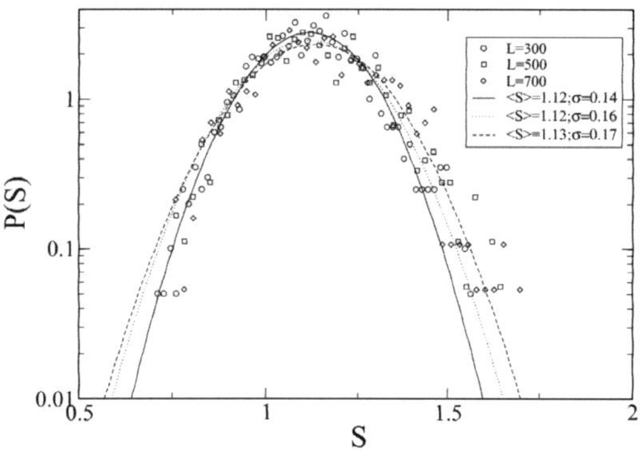

Figure 1. The distribution $P(S)$ of the EE S at the midpoint of the sample. The disorder was chosen as $W = 0.7$ ($\xi = 214$) and there are no electron-electron interactions ($U = 0$). Symbols depict the numerical results, while the curves correspond to a fit to the normal distribution with an average and variance given in the legend.

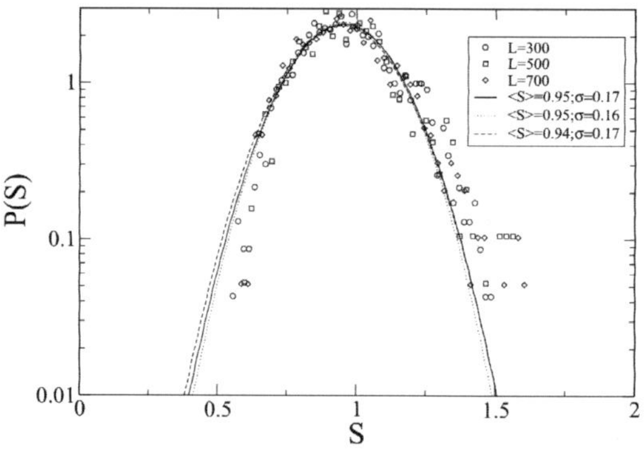

Figure 2. As in Fig 1, but in this figure the strength of the electron-electron interactions ($U = 0.6$) corresponds to a many particle localization length of $\xi = 57$. Thus, for all length of the sample we are in the strong localized regime. It can be clearly seen that the normal distribution gives a rather poor fit to the numerical results.

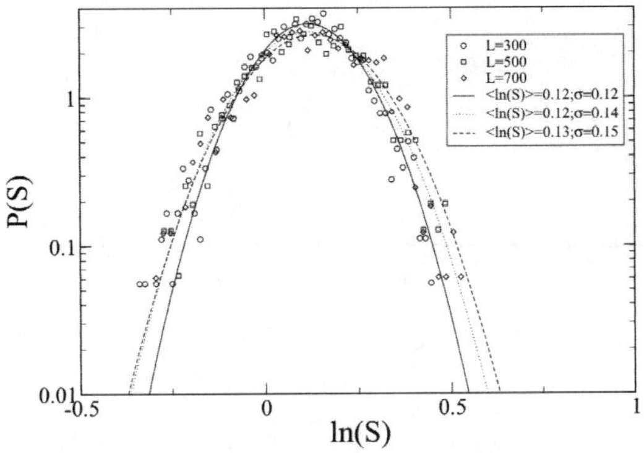

Figure 3. As in Fig 1 ($U = 0$), except here we fit the numerical distribution to a log-normal distribution.

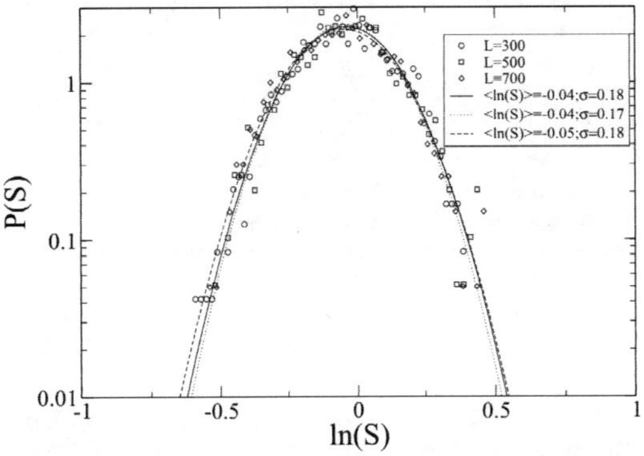

Figure 4. As in Fig 2($U = 0.6$), except here we fit the numerical distribution to a log-normal distribution.

First we examine the distribution of the EE for a non-interacting case ($U = 0, g = 1$). In all cases we plot the distribution of the EE at the middle of the sample, i.e., $L_A = L/2$. The results are depicted in Fig. 1. For the shortest samples, $L = 300$ (for which $\xi \approx L$) the fit to the normal distribution is not to bad. For longer sample length $L = 500, 700$ the fit becomes worse. When the interacting case is considered (See Fig. 2), for which $\xi(U = 0.6) = 57 \ll L$ for all length considered here, the fit badly fails, especially in the tail of the distribution.

Alternatively we fit the numerical data to a log-normal distribution as is shown in Figs. 3,4. While for the non-interacting case (Fig. 3) the log-normal fit is not great (indeed, it is quite bad for the shortest samples $L = 300$), it fits pretty well for the interacting case (Fig. 4). Thus it seems that the numerical data supports a crossover from the normal distribution at the weak localization regime ($L \leq \xi$) to a log-normal distribution at the strong localization limit ($L \gg \xi$).

4. Discussion

This crossover of the EE distribution could have some implications to the study of the EE in disordered systems. The most obvious is related to using the EE as an indicator of the position and properties of the localization transition. As we mentioned the EE is expected to saturate once $L_A > \xi$. The most natural approach would be to study the average value of the EE over different realizations of disorder. Nevertheless, once the distribution is not a normal one must be careful in performing the average. Since one would like to span the whole regime from $L_A \ll \xi$ to $L_A \gg \xi$ without changing the averaging method, probably the best approach would be to treat the median of the distribution as the representative EE for the system [5].

It is also interesting to try to extrapolate from the one-dimensional behavior of the EE to the general behavior of the EE at any dimension. Leaning on the similarity of the distribution of conductance and EE for one-dimensional systems it is tempting to speculate that also for higher dimensions the EE distribution will follow the normal distribution for the metallic regime (with a fixed width like the UCF?), while it should follow a log-normal distribution in the localized regime. These speculations need to be substantiated by further studies.

5. Acknowledgments

Financial support from the Israel Science Foundation (Grant 686/10) is gratefully acknowledged.

References

[1] For recent reviews see: Amico L, Fazio R, Osterloh L, and Vedral V 2008 *Rev. Mod. Phys.* **80**, 517; Eisert J, Cramer M, and Plenio M B 2010 *Rev. Mod. Phys.* **82**, 277; and references therein.
[2] Sachdev S 1999 *Quantum Phase Transitions* (Cambridge University Press, Cambridge).
[3] Vojta M 2006 *Phil. Mag.* **86**, 1807.
[4] Le Hur K 2008 *Ann. Phys.* **323**, 2208.
[5] Berkovits R 2011 *in preperation.*
[6] Goldstein M, Gefen Y and Berkovits R 2011 *Phys. Rev. B* **83**, 245112; Goldstein M, Berkovits R and Gefen Y 2010 *Phys. Rev. Lett.* **104**, 226805.
[7] Apel W 1982 *J. Phys. C* **15**, 1973.
[8] Woynarovich F and Eckle H P 1987 *J. Phys. A* **20**, L97; Hamer C J, Quispel G R W and Batchelor M T 1987 *ibid.* **20**, 5677.
[9] Giamarchi T 2003 *Quantum Physics in One Dimension* (Oxford University Press, New York).
[10] White S R 1992 *Phys. Rev. Lett.* **69**, 2863.
[11] Schollwöck U 2005 *Rev. Mod. Phys.* **77**, 259; Hallberg K A 2006 Adv. Phys. **55**, 477.
[12] Anderson P W, Thouless D J, Abrahams E and Fisher D S 1980 *Phys. Rev. B 22*, 3519; Altshuler B L, Kravtsov V E and Lerner I V 1986 *Zh.Eksp. Teor. Fiz.* **91**, 2276 [*Sov. Phys. JETP* **64**, 1352].

Journal of Physics: Conference Series **376** (2012) 012021 doi:10.1088/1742-6596/376/1/012021

Compositional disorder anomalies in Ga(N,P,As)/GaP quantum well structures

K Jandieri, S D Baranovskii, B Kunert, M Zimprich, S Liebich, K Volz, W Stolz, S Chatterjee, C Karcher, W Heimbrodt and F Gebhard

Department of physics and material sciences center, Philipps university Marburg, D-35032 Marburg, Germany

E-mail: kakhaber.jandieri@physik.uni-marburg.de

Abstract. Photoluminescence (PL) in many-component semiconductor heterostructures is strongly affected by disorder potential caused by compositional fluctuations. Our experimental study of the temperature-dependent PL in Ga(N,P,As)/GaP QW indicates the intriguing feature that the scale of disorder potential decreases with increasing concentration of the fluctuating compositional component (nitrogen). We claim that this effect strongly indicates that the dependence of the band structure on the concentration of the fluctuating compositional component weakens and the effective mass decreases with increasing concentration (in $2\% \div 4\%$ domain). This conclusion is supported by theoretical estimates within the analytical theory of compositional fluctuations in mixed crystals..

1. Introduction

The application of QW structures in optoelectronic devices requires detailed knowledge of their optical quality. Being grown at extreme non-equilibrium conditions the QW structures inevitably posses a certain degree of disorder due to their alloy structure and/or to imperfect interfaces. The disorder gives rise to carrier localization that dramatically affects optical properties. Measurements of the photoluminescence (PL) response have become a standard tool for characterizing the quality of the sample. In many situations, depending on various microscopic parameters, the Coulomb interaction between the photo-excited electrons and holes can be treated by considering excitons moving in an effective disorder potential [1]. Because of exciton localization, the PL peak energy and the PL linewidth demonstrate non-monotonous temperature dependencies shown schematically in Fig.1. The main experimentally observed trends of temperature-dependent PL spectra are (i) an anomalous redshift of the PL peak energy in the range of intermediate temperatures, so-called S-shape behavior; (ii) a peak of the PL linewidth within a narrow temperature range. Experimental study of the PL properties supplied by kinetic Monte-Carlo simulations gives valuable information about the energy scale of the disorder, and hence about the quality of the QW structure.

The experimentally observed effects of disorder on the dynamics of recombining carriers can be reproduced within a model where spatially localized excitons hop between localized states distributed randomly in space and energy [2]. The important input of the model is the density of states (DOS) in the band tail. The model can be applied to any kind of disordered systems,

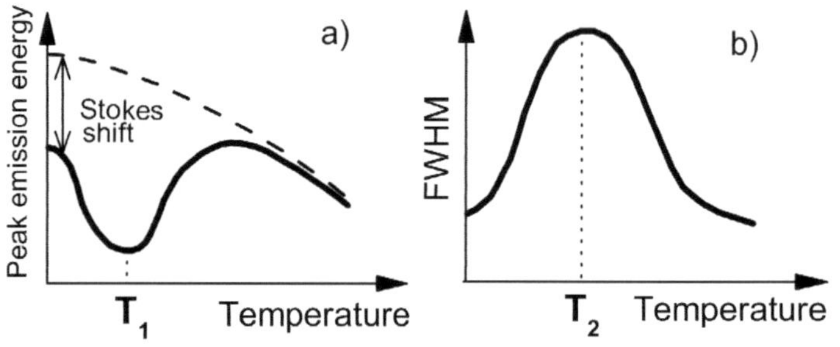

Figure 1. Schematic shapes of the temperature dependent a) PL peak position (solid line) and conduction band edge (dashed line) and b) PL linewidth, with taking into account the hopping transitions of carriers between the localized states.

though the results are rather sensitive to the shape and to the energy scale of the DOS. Therefore the correct determination of the DOS is the task of crucial importance. While single energy and spatial scales of disorder were necessary to explain experimental data in ZnCdSe/ZnSe, (GaIn)(NAs)/GaAs, (GaIn)(NAs)/GaNAs, Ga(NAs)/GaAs QW structures [1, 3], disorder with several scales was assumed to interpret experiments in (Al,In,Ga)N and Ga(As,Bi) material systems [4]. Another interesting example that evidences two, short- and long-ranged scales of disorder is Ga(N,P,As)/GaP QW structure recently studied in Ref. [5]. The long-range disorder was attributed to the variation of QW thickness due to the interface imperfections while the short-range disorder was attributed to the compositional fluctuations inherent in any compound material. Monte-Carlo simulations using this model reveal the exponential distribution of both, short-range and long-range disorders with characteristic energies E_0 and ε_0, respectively, as well as several universal relations giving the possibility to extract these energy scales from the experimentally observed PL features.

In the present paper we show another interesting and intriguing feature of Ga(N,P,As) compound. Intuitively one can suggest that the higher the nitrogen content in the compound gets, the high the energy scale of disorder should be since the absolute deviation of the composition from the average one increases with increasing concentration of the fluctuating compositional component (nitrogen in our case). Our experimental results evidence however the opposite effect: Both E_0 and ε_0 energy scales of the short-range and the long-range disorders decrease with increasing nitrogen concentration. We give a theoretical explanation of this "anomalous" effect using the analytical theory [1, 6] that allows one to determine the energy scale of compositional disorder from the structural and energy parameters of the compound. Further we focus our attention on the variation of short-range disorder since this kind of disorder is conditioned by the compositional fluctuations of the alloy [5]. The variation of long-range disorder is reserved for separate study.

Prior to presenting our experimental and theoretical results we briefly discuss the analytical theory of compositional disorder as well as the basics of the band-anticrossing (BAC) model [7] giving the compositional-dependent energy structure of the compound necessary for our theoretical considerations.

2. Analytical theory for compositional disorder

The short-range disorder in nitride III-V QW structures is conditioned by the compositional fluctuations inherent in the compound [5, 8]. Therefore, it is important to reveal the compositional dependent parameters determining the energy scale of the disorder potential. The corresponding analytical theory has been developed in Refs. [1, 6]. The basic idea of this theory is the following.

Let us assume that certain lattice sites may contain atoms of either type A or B. Let the concentration of such sites be M and the mean fraction of A sites be x. The mean number of A atoms in a volume $R \times R \times R$ is xMR^3, and the typical excess number is of the order of

$(xMR^3)^{1/2}$. The fluctuational change in the fraction of A atoms in this volume is then

$$\Delta x \simeq \frac{(xMR^3)^{1/2}}{MR^3}. \tag{1}$$

For small values Δx the compositional dependent energy $E_c(x)$ corresponding to the bottom of the conduction band may be expressed as

$$E_c \simeq E_c(x) + \alpha\Delta x, \tag{2}$$

where $\alpha = (\partial E_c/\partial x)_{\Delta x=0}$. Thus, in the considered volume, a potential well (or hump) is formed, with the depth (or height) of the order

$$V \simeq \alpha\frac{(xMR^3)^{1/2}}{MR^3}. \tag{3}$$

If V and R satisfy the inequality $V >> \hbar^2/mR^2$, where m is the effective electron mass, such a well has an electronic level close to V. If the opposite inequality is valid, the well contains no level at all. It is seen from Eq.(3) that the energy V increases with decreasing R. Consequently, the most effective of the wells having levels are those with small R. The most important wells are therefore those with the lowest value of R that generate a level [6]. This value is determined by the equation $V = \hbar^2/mR^2$, i.e.,

$$R \simeq \frac{\hbar^4 M}{\alpha x m^2}. \tag{4}$$

Substituting Eq.(4) in Eq.(3), we find that the smearing of the band edge is of the order of

$$E_0 \simeq \frac{\alpha^4 m^3 x^2 (1-x)^2}{\hbar^6 M^2}, \tag{5}$$

We use Eq. (5) to estimate the energy scale of compositional disorder E_0 in the samples under study. In our case the more precise definition of content x is necessary. We study the fluctuation disorder in quaternary Ga(N,P,As) compound where the content of both phosphorous and nitrogen components may fluctuate. However, the conduction band edge E_c is much more sensitive to the variation in the nitrogen rather than in the phosphorous content because of the extremely strong anticrossing interaction [7] between the localized nitrogen states and the extended states of the semiconductor matrix. Hence, $\alpha_N = \partial E_c/\partial x_N$ for nitrogen is much larger than $\alpha_P = \partial E_c/\partial x_P$ for phosphorous and concomitantly the disorder energy scale E_0 given by Eq. (5) is mainly determined by fluctuations of the nitrogen content. Therefore we imply under x the nitrogen content. Accordingly, M is the concentration of sites for group V elements in the compound.

The important input of Eq. (5) is the compositional dependence of the band gap in QW compound α. This parameter can be estimated from simple two-level band-anticrossing model [7]. The basics of this model are given in next Section.

3. Band-anticrossing (BAC) model

It is well known that even a small amount of incorporated nitrogen leads to a strong reduction of the band gap energy of III-V semiconductor compounds. In BAC model [7] the modification of the conduction band is described as an anticrossing interaction between highly localized states of the substitutional nitrogen atoms and the extended states of the host semiconductor matrix. According to this model the newly formed lowest E_c subband position can be estimated as

$$E_c(x) = \frac{1}{2}\left[(E_\Gamma + E_N) - \sqrt{(E_\Gamma - E_N)^2 + 4C^2 x}\right], \tag{6}$$

Table 1. Compositional and geometrical Characteristics of studied Ga(N,P,As)/GaP QW structures

Sample	X_N (%)	X_P (%)	X_{As} (%)	QW width (nm)
sample1	2	14	84	5.6
sample2	3	5	92	5.6
sample3	4	7	89	5.4

where E_Γ is the Γ conduction band edge in the host semiconductor matrix, E_N is the energy of the N level and C is the coupling parameter.

Originally Eq. (6) has been applied in order to parameterize the strong band gap bowing of GaAs or GaP-based material systems such as Ga(N,As), (Ga,In)(N,As) and Ga(N,P). Recently the BAC model has been adopted for As-rich Ga(N,P,As) mixed compounds [9] by extrapolation of E_N and E_Γ energy levels from Ga(N,As) and Ga(N,P) ternary endpoints. Furthermore, the modification of these energy levels as functions of the macroscopic strain inherent in QW structure was also taken into account. We use the adopted BAC model to estimate the band gap variation α in studied Ga(N,P,As)/GaP QW structures. The estimated values of this material characteristic are then exploit to determine the energy scale of compositional disorder E_0 according to Eq. (5).

This program is performed in Section 5 to provide the theoretical explanation of the experimentally observed anomalous trends in the compositional-dependent energy scale of disorder in Ga(N,P,As)/GaP QW structures.

4. Experimental results

We study the photoluminescence behavior of three Ga(N,P,As)/GaP QW structures with different compositional and geometrical characteristics gathered in Table 1. The experimental results presented in Fig. 2 show the following tendencies with increasing nitrogen content: (i) the linewidth of the PL spectra decreases, (ii) the temperatures corresponding to the minimal peak energy and to the maximal linewidth of the PL spectra decrease. According to Ref. [2], the temperatures corresponding to the minimal peak energy and to the maximal linewidth of the PL spectra (T_1 and T_2, respectively. See also Fig. 1a,b) are related with the energy scale of the short-range disorder E_0 by the following universal relations

$$\begin{aligned} k_B T_1 &= (0.75 \div 0.80)\, E_0, \\ k_B T_2 &= (1.10 \div 1.15)\, E_0. \end{aligned} \tag{7}$$

Eqs. (7) give the possibility to extract the energy scales of compositional disorder in studied Ga(N,P,As)/GaP QW structures from the experimentally observed PL features presented in Fig. 2. The corresponding nitrogen content dependence of the energy scale E_0 shown in Fig. 3a (open circles) evidences the significant reduction of the compositional disorder with the increase of nitrogen concentration resulting in decreasing linewidth of the PL spectra.

In the next Section we exploit the analytical theory of compositional disorder and the band-anticrossing model in order to estimate the energy scale of compositional disorder E_0 in Ga(N,P,As)/GaP QW structures with different nitrogen contents x and compare the obtained $E_0(x)$ dependence with that extracted from our experimental studies.

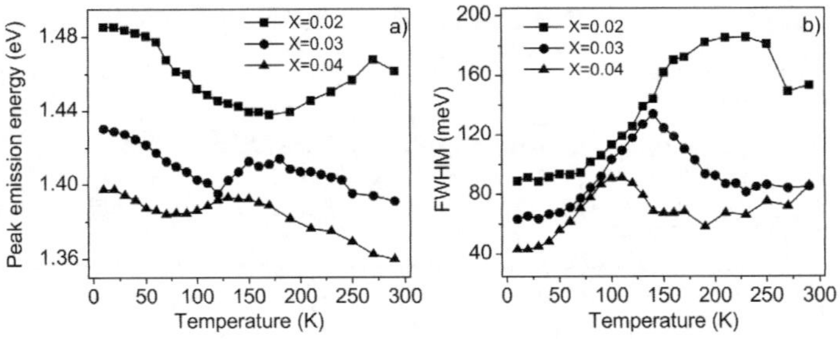

Figure 2. Temperature dependent a) peak energy and b) linewidth of PL emission spectra in Ga(N,P,As)/GaP QW structures with different nitrogen content x.

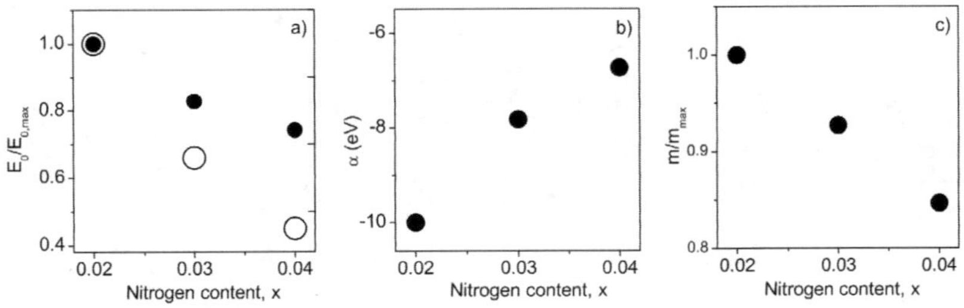

Figure 3. Nitrogen content dependencies of: a) the energy scale of compositional disorder extracted from the experimentally observed PL features (open circles) and calculated from Eq. (5) with constant effective mass (solid circles), b) the compositional variation of the band gap calculated according to BAC model, c) the effective mass determined from the fitting of the experimentally observed and the calculated energy scales of compositional disorder.

5. Calculation results and comparision with experiment

The analytical theory presented in Section 2 gives the possibility to relate the energy scale of compositional disorder with important material characteristics such as the composition-dependent band-structure and the effective mass. To determine the band-structure we use the BAC model (see Eq. (6) of Section 3) adopted to Ga(N,P,As) quaternary compounds [9]. The calculated nitrogen content dependencies of the variation of the conduction band edge α is presented in Figs. 3b. The resulting dependence of the relative E_0 scale of compositional disorder on the nitrogen content calculated according to Eq. (5) assuming the constant effective mass is shown in Fig. 3a (solid circles). The comparison of the calculated energy scales of compositional disorder with those extracted from PL features observed in our experimental studies (open circles in Fig. 3a) evidences that the analytical theory of compositional disorder gives the qualitative explanation of the observed anomalies in Ga(N,P,As)/GaP QW structure, i.e. the energy scale of disorder decreases with increasing nitrogen content. However, there still exists the quantitative discrepancy between the calculated and the experimental results. We account for this discrepancy with the assumption of the constant effective mass used in our calculations. The values of the effective mass are not known so far for Ga(N,P,As) material system. The situation is not completely clear even for Ga(N,As) and Ga(N,P) ternary compounds (see for example Ref. [10]). An analytical theory of compositional fluctuations [1, 6] opens the opportunity to determine the compositional variation of the effective mass in As-rich Ga(N,P,As) quaternary compounds. The relative values of the effective mass as a function of the nitrogen content (see Fig. 3c) are estimated as the values providing the exact agreement between the experimentally observed and the calculated energy scales of compositional disorder. According to our estimations, the effective mass decreases with increasing nitrogen concentration, which is a common feature of dilute III-V nitride compounds with comparatively high nitrogen content [10, 11].

6. Conclusions

The experimental investigation of the photoluminescence behavior in Ga(N,P,As)/GaP QW structures reveals the anomalous compositional dependence of the disorder parameters. In particular, the energy scale of disorder decreases with increasing nitrogen content.

Using the theory of compositional fluctuations in mixed crystals it is shown that the reason for reducing disorder is the reducing effective mass and the specific dependence of the conduction band edge on the nitrogen content in Ga(N,P,As)-material systems. This dependence becomes weaker with increasing nitrogen content.

Acknowledgments

Financial support of the Fonds der Chemischen Industrie and that of the Deutsche Forschungsgemeinschaft is gratefully acknowledged.

References

[1] Baranovskii S D and Efros A L 1978 *Soviet physics: Semiconductors* **12** 1328

[2] Baranovskii S D, Eichmann R and Thomas P 1998 *Phys. Rev.* B **58** 13081

[3] Niebling T, Rubel O, Heimbrodt W, Stolz W, Baranovskii S D, Klar P J and Geisz J F 2008 *J. Phys.* C **20** 015217

Rubel O, Baranovskii S D, Hantke K, Kunert B, Rüle W W, Thomas P, Volz K and Stolz W 2006 *Phys. Rev.* B **73** 233201

Rubel O, Galluppi M, Baranovskii S D, Volz K, Geelhaar L, Riechert H, Thomas P and Stolz W 2005 *J. Appl. Phys.* **98** 063518

Grüning H *et al.* 2004 *Phys. Status Solidi* C **1** 109

[4] Zukauskas A, Kazlauskas K, Tamulaitis G, Khan M A, Yang J W, Zhang J, Simin G, Shur M S and Gaska R 2003 *Phys. Status Solidi* C **0** 2737

Kazlauskas K, Tamulaitis G, Zukauskas A, Khan M A, Yang J W, Zhang J, Simin G, Shur M S and Gaska R 2003 *Appl. Phys. Letters* **83** 3722

Kazlauskas K, Tamulaitis G, Pobedinskas P, Zukauskas A, Springis M, Huang Chi-Feng, Cheng Yung-Chen and Yang C C 2005 *Phys. Rev.* B **71** 085306

Imhof S *et al.* 2010 *Appl. Phys. Lett.* **96** 131115

[5] Karcher C, Jandieri K, Kunert B, Fritz R, Zimprich M, Volz K, Stolz W, Gebhard F, Baranovskii S D and Heimbrodt W 2010 *Phys. Rev.* B **82** 245309

[6] Alferov Zh I, Portnoi E L and Rogachev A A 1969 *Soviet Physics: Semiconductors* **2** 1001

[7] Shan W, Walukiewicz W, Ager III J W, Haller E E, Geisz J F, Friedman D J, Olson J M and Kurtz S R 1999 *Phys. Rev. Lett.* **82** 1221

Shan W, Walukiewicz W, Yu K M, Ager III J W, Haller E E, Xin H P and Tu C W 2000 *Appl. Phys. Lett.* **76** 3251

Wu J, Shan W and Walukiewicz W 2002 *Semicd. Sci. Technol.* **76** 860

[8] Jandieri K, Jurecka C, Ohlmann J, Beyer A, Kunert B, Baranovskii S D, Volz K, Stolz W and Gebhard F 2011 *Phys. Status Solidi* C **8** 163

[9] Kunert B, Volz K, Koch J and Stolz W 2006 *Appl. Phys. Lett.* **88** 182108

[10] Dannecker T, Jin Y, Cheng H, Gorman C F, Buckeridge J, Uher C, Fahy S, Kurdak C and Goldman R S 2010 *Phys. Rev.* B**82** 12203

[11] Masia F *et al* 2006 *Phys. Rev.* B **73** 073201

Journal of Physics: Conference Series **376** (2012) 012022 doi:10.1088/1742-6596/376/1/012022

Calculation of charge carrier concentration profile and magnetoresistance in modulation doped structures with a wide potential well

L. Shchurova

Lebedev Physical Institute of Russian Academy of Science, Moscow, Russia

E-mail: ljusia@gmail.com

Abstract. We investigated Fermi charge carriers' concentration profile in modulation doped structures with a wide quantum well. We have formulated hydrodynamic equations both in the absence of magnetic field and in parallel magnetic field. We have obtained analytical solutions for the carrier concentration as a function of the tangential coordinate in the potential well. In a quantum area near the interface, we carried out quantum mechanical calculations taking into account the effect of electric and magnetic fields. The carrier concentration profile and magnetoresistance in parallel magnetic field are presented for modulation doped Si/SiGe/Si structures.

1. Introduction

We investigate the properties of the charge carrier systems with the Fermi statistics in modulation doped structures with a wide quantum well. All carriers in the two-dimensional conducting channel are formed due to the ionization of impurities in the doped layer, separated from the quantum well by the spacer. We study the carrier system with a sufficiently high density, such that effectively three-dimensional inhomogeneous electron state is formed. The carrier concentration profile is asymmetric due to significant interaction of the electrons with the electric field of the doped layer. In such structures the calculations of the Fermi electron state becomes difficult to calculate since screening effects are to be taken into account.

The problem of calculating such carrier concentration profile for self-consistent confinement potential has a long history. It was solved numerically many times (in particular, in ref. [1]). Approximate analytical solutions were only performed with some simplifying assumptions (in particular, in ref. [2]). The screening was assumed to be much smaller than the well width, which is satisfied only for very high doping level. However, carrier systems with the screening length comparable to the well width are of interest because of greater mobility and significant effects of magnetoresistance [3].

Calculations of Fermi carriers' states in modulation doped structures in strong magnetic fields require consideration of the self-consistent effect of the screening and the magnetic field, so, the problem becomes even more complicated.

This paper is devoted to derivation of the charge carrier concentration profile model in modulation doped structures both in the absence of magnetic field and in parallel magnetic field. The point of our work is to consider the screening effect without further simplifying assumptions. We have formulated a hydrodynamic model and obtained an analytical solution for the charge carrier concentration as a function of the transverse coordinate in the quantum well. In a quantum area (near the interface) we solved the Schrodinger equations.

2. Calculation of carriers' concentration profile in a potential well without magnetic field

In this paper, we study the carrier system with a sufficiently high density a wide quantum well. In such system, the de Broglie wavelength of electrons is smaller than the potential well width. So the characteristic length of change of the wave function (the length of the de Broglie wavelength λ) is smaller than the characteristic length of potential variation (quantum well width *L*). In this case an effectively three-dimensional inhomogeneous electron state is formed. This electron system is quasi-classical (except for regions near the interface) and the hydrodynamic approximation can be used.

Within the hydrodynamic approach, we calculated a three-dimensional charge carrier concentration $n(z)$, which changes only in transversal coordinate *z*, on the basis of the thermodynamic condition of total energy conservation:

$$\mu + E_{\text{int}} = const \tag{1}$$

Here $\mu = \mu(n(z)) = \frac{(3\pi^2)^{2/3}}{2}\frac{\hbar^2}{m_h^{2/3}} n(z)^{2/3}$ is the chemical potential of the charge carrier system, $E_{\text{int}} = E_{\text{int}}(n(z)) = e \cdot \Phi(z)$ is the electrostatic interaction energy. The electrostatic potential Φ is determined by the Poisson equation $\frac{d^2\Phi}{dz^2} = \frac{4\pi \cdot e}{\chi} n(z)$, where χ is the permittivity of the medium.

We are considering structures with the impurity layer is located far enough from the quantum well, so that the distance between the ions and charge carriers is far greater than the distance between the ions in the doped layer. In this case we can assume that charge carriers interact with the whole ions' layer. Then the Poisson equation implies

$$e \cdot \frac{d\Phi}{dz} = e \cdot \left(E_i - \frac{4\pi \cdot e}{\chi} \cdot \int_0^z n(\zeta) \cdot d\zeta \right) \tag{2}$$

Here, $z=0$ is a coordinate of the interface from the side of the doped layer and E_i is the unscreened field of the ions' layer.

Differentiating (1) and taking into account (2), we get the equation for $n(z)$:

$$-E_0 + K_0 \int_0^z n(\zeta) \cdot d\zeta = [n(z)]^{-1/3} \cdot (dn/dz). \tag{3}$$

Here $E_0 = \frac{2 \cdot e \cdot m \cdot E_i}{3(3\pi)^{2/3} \cdot \hbar^2}$, $K_0 = \frac{8\pi \cdot e^2 \cdot m}{3(3\pi)^{2/3} \cdot \hbar^2 \cdot \chi}$.

The equation (3) describes the equilibrium of attractive forces of electrons to the ions' layer (the left-hand side of the equation) and the repelling forces of electrons by the Fermi pressure (the right-hand side of the equation) in each interval *dz* in the potential well.

We solved (3) with the boundary condition $dn/dz|_{z=L} = 0$, which means that on the right boundary of the potential well the ions' field is fully offset by the field of charge carriers. We have obtained the analytical solution of the equation (3) in terms of hypergeometric functions:

$$z(n) = L - K \cdot n_L^{-5/6} \cdot (n^{2/3}) \cdot F\left(\frac{2}{5}, \frac{1}{2}, \frac{7}{5}, \sqrt{(n/n_L)^{5/3} - 1}\right) \tag{4}$$

with $K = \frac{3}{8}\cdot\left(\frac{5\cdot(3\pi)^{2/3}\cdot\hbar^2\cdot\chi}{8\pi\cdot e^2\cdot m}\right)^{1/2}$ and $n_L = n|_{z=L}$. Then $n(z)$ is given by the inverse function to $z(n)$.

It is remarkable that our solution is valid for any ratio between the screening length and well width. Detailed solution of the integro-differential equation (3) is given in ref. [4].

Calculation by formula (4) gives a sharp peak for $n(z)$ in the area immediately adjacent to the barrier (Figure 1).In this area, the hydrodynamic model is no longer applicable. Due to a rapid increase of Fermi energy, Fermi wave vector k_F increases faster than z, i.e. $dk_F/k_F \geq dz/2\pi z$. So, there are not enough particles with a definite wavelength $dz/2\pi$.The intersection of curves in the inset of Figure 1 shows the coordinate $z \geq z_B$ where hydrodynamic model starts to be valid.

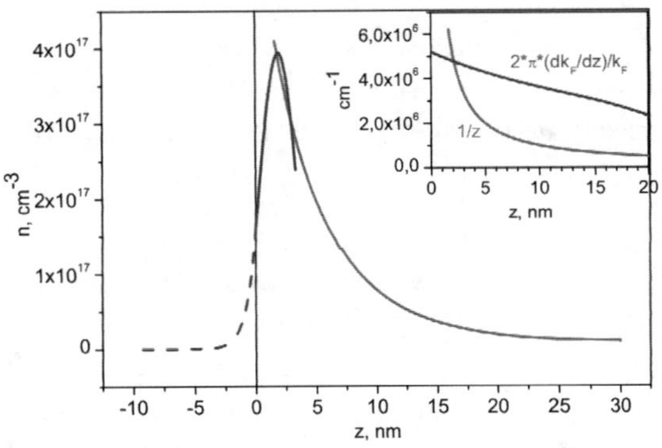

Figure 1. Concentration profile of holes in Si/SiGe/Si structure with doping level $N_i=2.2\cdot10^{11}\ cm^{-2}$, $L=30\ nm$, $U_B=92\ meV$. Solid red line is calculations using hydrodynamic model. Blue line is computed using Schrodinger equation in mean-field approximation.

In quantum area we have to solve the effective Schrodinger equation

$$-\frac{\hbar^2}{2m}\frac{d^2\Psi(z)}{dz^2} + V(z)\cdot\Psi(z) = E\cdot\Psi(z)$$

We have used the electron wave in the well as $\Psi_W(z) = C_0 Ai(x/l - \delta)$ and in the barrier as $\Psi_B(z) = A_0\exp(k_b z)$, where $l = (\hbar^2/2meE_i)^{1/3}$, k_B is the wave vector in the barrier. The coefficients A_0 and δ were calculated from the continuity condition of wave function and its derivative at z=0. The coefficient C_0 and the value of n_L were chosen so that $\int_{L_p}^{L} n(z)dz = N_i$, where N_i is a two-dimensional density of ions, L_P is depth of penetration into the barrier, $L_P<0$.

We described the quasi-classical area using the hydrodynamic approach and quantum area using the Schrodinger equation. The results of our calculations according to the above model are presented in Figure 2. This profile differs substantially from the symmetrical concentration profile of two-dimensional charge carrier system in a quantum well with quantum confinement levels.

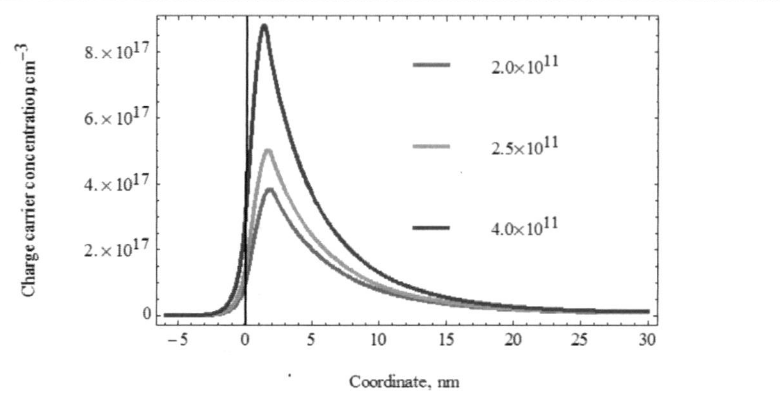

Figure 2. Holes'concentration profile in Si/SiGe/Si modulation doped structure with doping level $N_i = 2\cdot 10^{11}$, $2.5\cdot 10^{11}$ and $4\cdot 10^{11}$ cm^{-2}, $L=30$ *nm*.

3. Calculation of carriers' concentration profile in a potential well in parallel magnetic field

Now we study the charge carrier system in a magnetic field H applied parallel to the two-dimensional layers (along the x-axis). We consider the Fermi system with $\lambda < L$. Moreover, we study charge carrier system that satisfies the condition $k_F \cdot \lambda_H > 1$, where $\lambda_H = (\hbar c / eH)^{1/2}$ is the magnetic length. In this case the tlectron wave function is formed at the scale smaller than the magnetic length. In this case the hydrodynamic approach is applicable. We neglect spin-related effects, because the Zeeman spin splitting in a parallel field is small (in p-Si/SiGe/Si quantum well) [5].

In an external magnetic field, in addition to electrical forces, Lorentz force $\vec{F} = \frac{e}{c}\left[\vec{v} \times \vec{H}\right]$ acts on the charge carriers. Since the magnetic field is directed along the x-axis and the transversal velocity is zero in the equilibrium state, then the Lorentz force has only z-component: $F_L = -\frac{e}{c} v_y H_x$. The charge carriers in a magnetic field have a local velocity $v_y(z) = \hbar^{-1} \cdot \partial\varepsilon_k(z)/\partial k$, where $\varepsilon_k(z)$ is the total energy of the charge carriers. The velocity v_y has different signs near the opposite interfaces (according to the slope of the charge carrier profile). There are local currents $\vec{j}_y$ in the quantum well. (Currents lines are closed and the resulting drift current in the y-direction is zero.)

The Fermi carrier system possesses diamagnetic properties. The diamagnetic susceptibility χ_{dia} is given by $\chi_{dia}(z) = -\frac{1}{3} \cdot \frac{e^2 \cdot 3^{1/3} \cdot \pi^{-4/3}}{4 \cdot m \cdot c^2} \cdot n^{1/3}(z)$. An induced magnetic field H_{ind} arises due to diamagnetic effect: $H_{ind}(z) = 4\pi \cdot \chi_{dia}(z) \cdot H$. The total magnetic field $\vec{H}_{tot}(z) = \vec{H} + \vec{H}_{ind}(z)$ is the sum of external field and the induced field. From the Maxwell equation

$$rot\,\vec{H}_{tot} = \frac{4\pi}{c}\vec{J} + \frac{1}{c}\frac{\partial \vec{E}}{\partial t},$$

where $\vec{J}$ is the total current, and time derivative is zero (because we consider the steady state), we get

$$\frac{\partial}{\partial z} H_{ind}(z) = \frac{4\pi}{c} J_y(z)$$

Substituting $J_y(z) = e \cdot n(z) \cdot v_y(z)$ we obtain $v_y(z) = -\frac{1}{12} \cdot \frac{e}{c} \cdot \frac{3^{1/3} \cdot \pi^{-4/3}}{m} \cdot H_x \cdot \frac{1}{3} \cdot n^{-5/3}(z) \cdot \frac{\partial n(z)}{\partial z}$.

Then the force balance equation takes the form

$$-E_0 + K_0 \cdot \int_0^z n(\zeta)\cdot d\zeta = \left\{(n(z))^{-1/3} + K_1 \cdot (n(z))^{-5/3}\right\}\frac{d}{dz}(n(z)). \tag{5}$$

Here $K_1 = \dfrac{e^2}{c^2}\dfrac{H_x^{\ 2}}{54\cdot 3^{1/3}\cdot \pi^{8/3}\hbar^2}$. This equation describes the equilibrium of attractive forces of electrons to the ions' layer (the left-hand side of the equation) and Lorentz force and the repelling forces of electrons by the Fermi pressure (the right-hand side) in each interval dz in the potential well.

In a magnetic field, the Fermi gas pressure changes due to the diamagnetic effect. Although magnitude of the induced magnetic field $H_{ind}(z)$ is quite small, the effect depends upon the derivative of total field $H + H_{ind}$, and thus the influence of diamagnetic effect becomes substantial. The changing total magnetic field is related to the local current $j_y(z)$ through the Maxwell equation and therefore it is related to the pressure of carriers' gas. This current flows orthogonal to the magnetic field and also orthogonal to the pressure gradient, and creates additional pressure.

We solved (5) with the boundary conditions $dn/dz\big|_{z=L} = 0$. While the integro-differential equation (5) is quite complicated we managed to solve it analytically

$$z(n) = L - 3\cdot\left(\frac{6}{5}K_0\right)^{-1/2}\left\{ n_L^{-1/6}\cdot \int_1^{(n/nL)^{1/3}} \frac{y\cdot dy}{\sqrt{(y^5-1)+\dfrac{5\cdot K_1}{n_L^{4/3}}(y-1)}} + K_1\cdot n_L^{-3/2}\int_1^{(n/nL)^{1/3}} \frac{y^{-3}\cdot dy}{\sqrt{(y^5-1)+\dfrac{5\cdot K_1}{n_L^{4/3}}(y-1)}}\right\}$$

where the n_L is the charge carrier density on the right interface.

Then $n(z)$ is given by the corresponding inverse function to $z(n)$. It is remarkable that our solution is valid for any ratio between the screening length and well width.

In this way we have obtained formula that describes the descending branch of the charge carrier concentration profile in the quantum well in parallel magnetic field (see Figure 3).

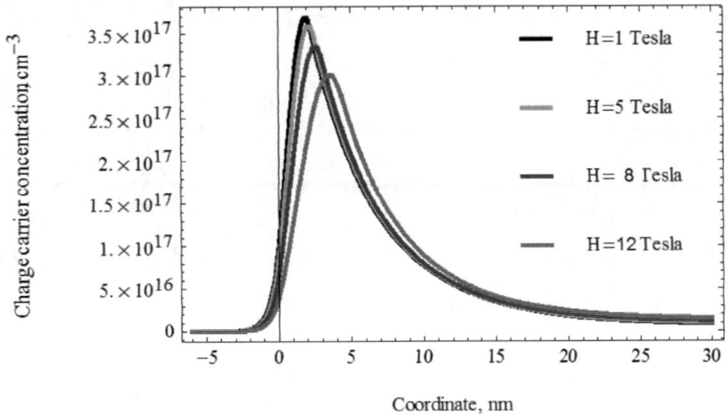

Figure 3. Holes'concentration profile in Si/SiGe/Si structure with doping level N_i=2.2·10^{11} cm^{-2}, L=30 nm, the parallel magnetic fields H=1, 5, 8, 12 Tesla.

In quantum area (near the interface) we have to solve the effective Schrodinger equation numerically taking into account the effect of electric and magnetic fields. Figure 3 shows our calculations of the profile of carriers' concentrations in the modulation doped Si/SiGe/Si structures. The concentrations'

Journal of Physics: Conference Series **376** (2012) 012022 doi:10.1088/1742-6596/376/1/012022

peak degrades with a magnetic field. The carrier density decreases in the region of the maximum. The rotation of the charge carriers under the magnetic field leads to “mixing” of carriers and redistribution of the carriers density in the quantum well.

Let us briefly discuss the magnetoresistance in a parallel magnetic field in modulation doped structures. In structures with a low charge carrier concentrations the giant magnetoresistance is associated with an orbital effect in a strong parallel magnetic field [6]. In the above structures with a sufficiently high doping level the carriers are located mainly near the interface in a region with a width of less than two Bohr radii ($r_c \sim 10^{-5} \div 10^{-6}\, cm$), thus the effect of orbital motion is weak for them. (Such a large value r_c is connected with high average density of particles in the peak concentrations.) Positive magnetoresistance is causedby by a decrease in the number of particles, which give a defining contribution to the drift current, and also a decrease in its average velocity (carriers’ mobility). Figure 4 shows the calculation of the resistivity ρ_{xx} in the electric field E_{appl} applied along the x-axis (parallel to the magnetic field) for the *Si/SiGe/Si* structure with doping level $N_i=2.2\cdot 10^{11}\ cm^{-2}$. It also shows the observed magnetoresistance from the ref. [3].

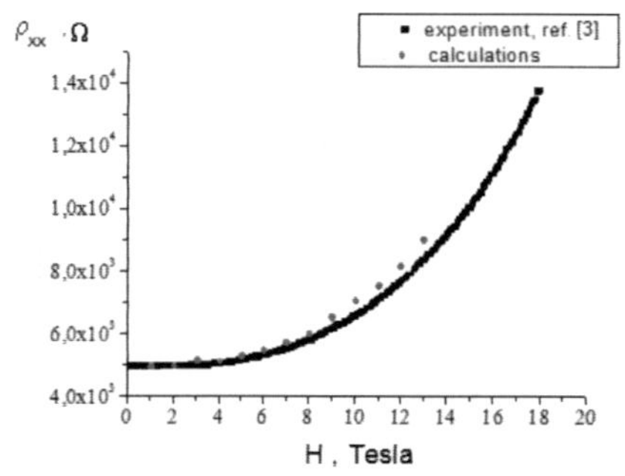

Figure 4. Magnetoresistance in a parallel magnetic field (E_{appl} II H). The black line is the experiment [3]. The red circles are our calculations

4. Conclusions

We have formulated hydrodynamic model and obtained analytical solutions for Fermi electrons’ concentration of as a function of tangential coordinate in the potential well and applied parallel magnetic field. These solutions are valid for any ratio between the screening length and well width. In a quantum area near the interface we carried out quantum mechanical calculations. The charge carrier concentration profile is significantly asymmetric, but becomes flatter in the strong magnetic field. Magnetoresistance is due to the carriers’ redistribution along the quantum well width with magnetic field.

References

[1] Tan I, Snider G, Chang L and Hu E 1990 *J. Appl. Phys.* **68** 4071
[2] Neumann F, Genenko Y, Melzer C, Yampolskii S and von Seggern H 2007 *Phys. Rev.* B **75** 205322
[3] Drichko I, Smirnov I, Suslov A, Mironov O and Leadley D 2009 *Phys. Rev.* B **79** 205310
[4] Shchurova L 2009 *Annalen der Physik* **18** 928
[5] Abrahams E, Kravchenko S and Sarachik M 2001 *Rev. Mod. Phys.* **73** 251
[6] Das Sarma S and Hwang E 2000 *Phys. Rev. Lett.* **84** 5596

Journal of Physics: Conference Series **376** (2012) 012023 doi:10.1088/1742-6596/376/1/012023

Resonant Tunneling in a Polymer with Charge Injection Doping

A. Ionov[a], R. Rentzsch[b]

[a] *Ioffe Physico-Technical Institute, Polytechnicheskaya 26, 194021 St. Petersburg, Russia*
[b] *Institut für Experimentalphysik, Freie Universität Berlin, Arnimallee 14, 14195 Berlin, Germany*

Abstract. It is shown that a supercurrent in disordered polymers placed between superconductor electrodes may be explained by resonant tunneling of Coopers pairs along polymer channels formed due to electrification effect. Polymer film thickness where supercurrent one may to observe depends on superconducting coherence length in superconductor. The more coherence length in superconductor electrode the more is a polymer film thickness where superconductor exists. Shapiro steps induced by microwave (f = 9.3 GHz) radiation correspond to a single Josephson junction.

Keywords: disordered polymer systems, electrification, supercurrent, resonant tunneling, Shapiro steps.

Introduction

It is well known that all insulators and non-conducting polymers always accept a charge when they are in contact with metals. It has been shown that polymer electrification occurs due to the localized traps situated in the polymer in an energy interval close to the Fermi energy of the majority of the metals [1, 2]. The traps are the result of various kinds of defects, such as chemical groups, cavities and entanglements, unsaturated bonds. The defects can appear in the polymer due to aging with time or they can be specially created by UV irradiation and oxidation [3]. The experiment shows that the appearance of metallic conductivity in metal-polymers-metal sandwiches and a supercurrent in superconductor-polymer-superconductor sandwiches is inevitably linked to this electrification process [4,5 and references therein]. The thermal motion of the molecules also contributes to the generation of a static charge, as well as other mechanical action (i.e. contact movement) increases the electrification. This explains the switching of metal-polymer metal structures to a highly conductive state under pressure or the type of contacting – when touching or sliding contacts were placed between the metallic electrode and the polymer surface. Electrification of polymers in metal-polymer-metal constructions as well as the arising metallic conductive state is not possible if the polymer has a permanent conductivity, for example due to initial doping. In "perfect" polymers without any defects the electrification penetrates only into the polymer surface, up to about tens of nm. For that reason there is no metallic conductivity in a polymer with higher film thickness.

In accordance with [6] one mechanism for the switching of the polymer from the dielectric state to the high conductive state can be the following. Polarization of the polymer matrix by randomly distributed localized charges causes both a shift and a broadening of the orbital energies, so that some HOMO-levels may lie above the Fermi level of the contact metal and some LUMO - levels below it. This means that the contact between a metal and a polymer causes electrons from the metal to transfer to the LUMO, or electrons may transfer from the HOMO into the metal for separated polymer macromolecules which are closely located near the surface of metallic electrodes [6]. As a result, energy states with energies close to the Fermi energy of most metals form between HOMO and LUMO orbital of macromolecules. Only weak van der Waals forces hold the macromolecules together in the solid. In the case of a small distance r between the macromolecules neighboring bulk macromolecules can except a charge and also may create the spreading of the energy states around the Fermi level of the metal. That mechanism can produce percolation paths of conductivity from one electrode to the other. In a disordered polymer system the charge transport between separate macromolecules or macromolecules clusters may exist due to resonant tunneling[1]. Early it was experimentally shown that the depth of charge penetration due to electrification may reach several micrometers [1]. We consider resonant tunneling as main process for the charge transport over large distances in polymers with randomly distributed in space internal defects.

To test the resonant tunneling model we investigated the critical current I_c behavior, current-voltage characteristics (CVC) of Superconductor-Polymer-Superconductor (SPS) junctions under different uniaxial pressure as well as with irradiation with microwave radiation at low temperature. As polymers we used imid-siloxane, polyimide and polyamidine [8]. All investigated polymers were with defects.

In accordance with [9] resonance transport of Cooper pairs along the paths made up of periodically arranged impurity atoms results in a slower decrease of the critical current with increasing barrier thickness then occurs at direct tunneling. More over the action of localized states act as long metallic channels so that the junction is similar to a long superconductor-normal metal-superconductor (SNS) junction. In the frame of that model we may use the theory of the proximity effect by De Gennes. De Gennes attributed the proximity effect to Cooper pairs that propagate from a superconductor into normal metal. The critical current I_c for a SNS junction in that approach is the following [10]:

$$I_c \cong \frac{\pi}{2eR_n}\frac{\Delta_i^2}{kT_c}\frac{d}{\xi_n}e^{-d/\xi_n} \quad (1)$$

Here are: I_c – critical current of weak link (normal metal N), R_n – resistance in normal state of N; Δ_i – superconducting gap at the normal interlayer interface, T_c – critical temperature of superconductor, d – the thickness of N, l_n – the mean-free path in the normal metal, v_n – Fermi velocity in N, ξ_n – the superconducting coherence length in N.

The temperature dependence of I_c results from the temperature dependence of ξ_n. For normal metal exist two limiting cases:

i) "clean limit" $\xi_s << l_n$ (here ξ_s – the superconducting coherence length in superconductor,). In this case one get $\xi_{n\ clean} = \hbar v_n / 2\pi kT$

ii) "dirty limit" $\xi_s >> l_n$

In this case $\xi_{n\ dirty} = (\hbar v_n l_n / 6\pi kT)^{1/2}$

The main predictions of the De Gennes theory are the following:

[1] On the possibility of the resonant tunneling in Superconductor-Polymer-Superconductor sandwiches was firstly discussed in [7].

Journal of Physics: Conference Series **376** (2012) 012023 doi:10.1088/1742-6596/376/1/012023

1) with increasing d of the normal metal, I_c should decrease exponentially at fixed ξ_n;

2) since ξ_n depends on T, I_c should exponentially decrease with increasing T up to T_c at fixed d;

3) I_c does not depend directly on ξ_s of the superconductor.

Before we proved experimentally the theoretical predictions for a SPS sandwich it should be noted the main dissimilarity to SNS which is that a polymer has no free carries as a normal metal. Hence the main problem of polymer in SPS construction with supercurrent is the value of superconducting coherence length (ξ_p) as well as its temperature dependence. When the LUMO of all macromolecules participate in the conductive paths from one electrode to another is aligned to the Fermi level then resonant tunneling of Coppers pars can exist in SPS. In this case the decay length of the superconductivity coherence length in polymers should be the same as in the superconductor electrodes, i.e. $\xi_p = \xi_s$. In this case ξ_n should be replaced by ξ_s in Eq. 1 and with decreasing ξ_s I_c should decrease exponentially in SPS at fixed polymer film thickness. The fundamental difference from normal metal is that I_c of a SPS junction should depend on the type of superconductor.

Experimental results and discussion

The dependence I_c versus film thickness d (nm) for imid-siloxane is presented in Fig.1 at fixed T and ξ_p. As seen from the figure the linear dependence of log I_c against d is in accordance with prediction Eq. 1. With increasing d up to some critical thickness d_c (d_c< 5-10 ξ_s) I_c decay to zero at $T<T_c$. In this case for polymer thickness up to 2 microns we observed a constant resistance which did not depend on T in a range between 1.8-3.4 K.

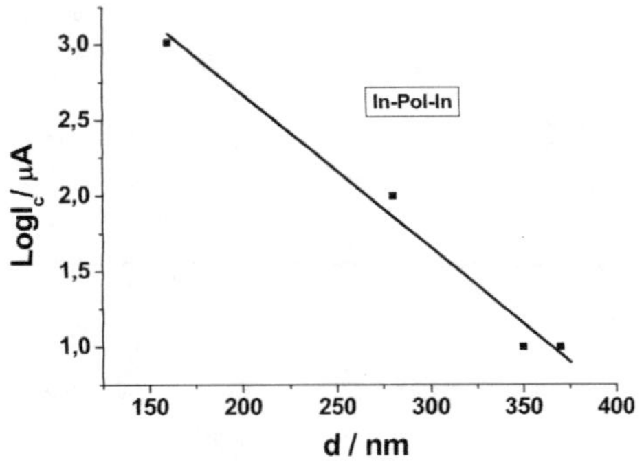

FIGURE 1. LogI_c dependence versus polymer film thickness d for In-Imid-Siloxane-In at fixed T_c close to T_c = 3.4 K. Contact areas between the polymer and superconductor are about the same.

Figure 2 presents the dependence R versus film-thickness for In-Polymer-In constructions at T = 4,2 K. One can see the linear dependence log$R$ against d (nm). This behavior is also in agreement with the resonant tunneling prediction where $R \infty \exp(-r/a)$, here r is a distance between localized states and a is the radius of localization. It should be noted that the result for the R(d) dependence obtained for imide-siloxane at low temperature differs from that obtained for atactic polypropelene at room temperature where R did not depend on d [3]. Possibly the reason is connected with the much less value of the spreading resistance between the polymer channel and the metallic electrodes in our case. The spreading resistance decreased down to about 2 orders of magnitude in comparison to room temperature. This increased the measurement accuracy of the polymer resistance.

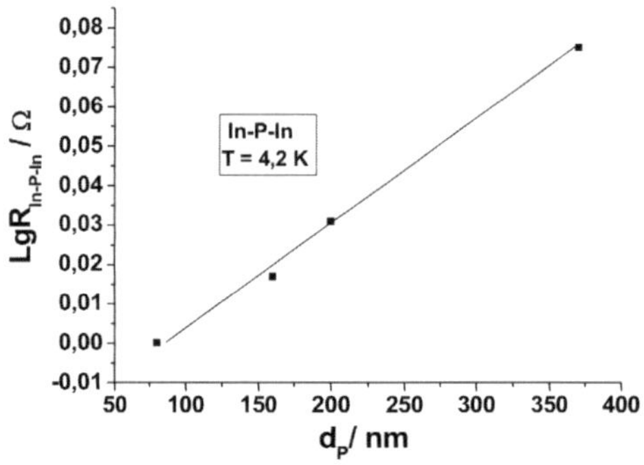

FIGURE 2. The dependence R versus film thickness d for In-P-In (P - imid-siloxane) sandwiches at fixed $T > T_c$.

Figure 3 shows the I_c dependence versus superconducting coherence length of such superconductors as In, Sn, Nb in SPS constructions. As seen from Fig.3 I_c obeys the Eq. 1 with $\xi_p = \xi_s$.

It follows as consequence that the detection of a supercurrent in highT_c-Polymer-highT_c constructions would be impossible with polymer film thickness more then some nm because of the small value ($\xi_{highTc} \sim 2$ nm).

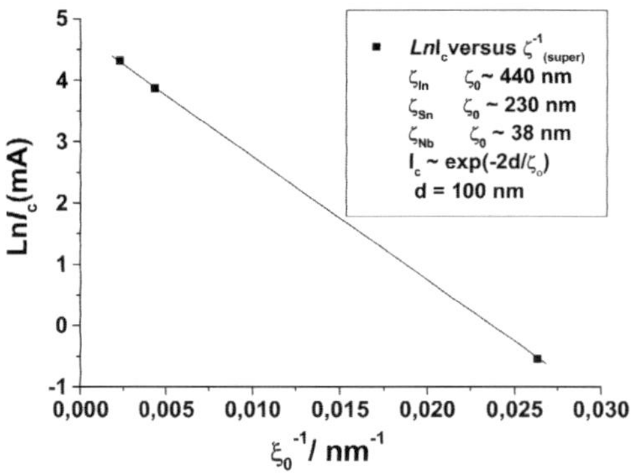

FIGURE 3. Dependence LnI_c versus the inverse superconducting coherence length of some superconductors.

Figure 4 shows the CVC of Josephson type for the flexible polymer polyamidine at small external pressure. As seen from the figure the CVC has a large hysteresis (indicated by the arrows). In accordance with resistively and capacitive shunted junction model the SPS junction with hysteresis loop has an essential capacitance [11]. The value of capacitance can be connected with the density of macromolecular chains in the polymer. The lower density of

Journal of Physics: Conference Series **376** (2012) 012023 doi:10.1088/1742-6596/376/1/012023

macromolecular packing the larger is a distance between macromolecules which participate in the charge transport.

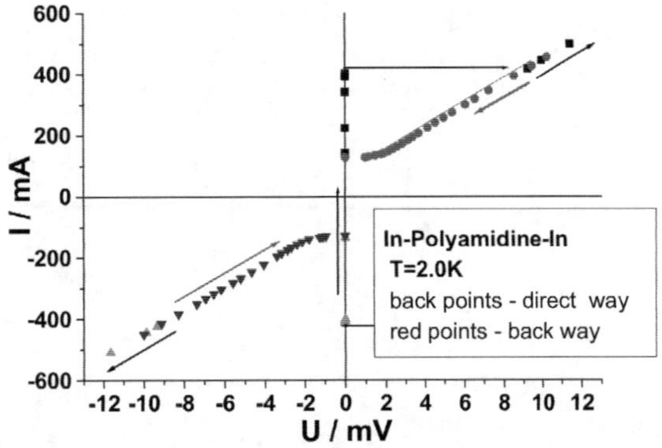

FIGURE 4. CVC of a S-P-S structure with high capacitance.

Figure 5 shows the dependence of the critical current I_c versus T. The saturation I_c at low temperatures is typical for a supercurrent transfer across a Josephson tunnel junction containing resonant localized impurity levels in the tunnel barrier [12].

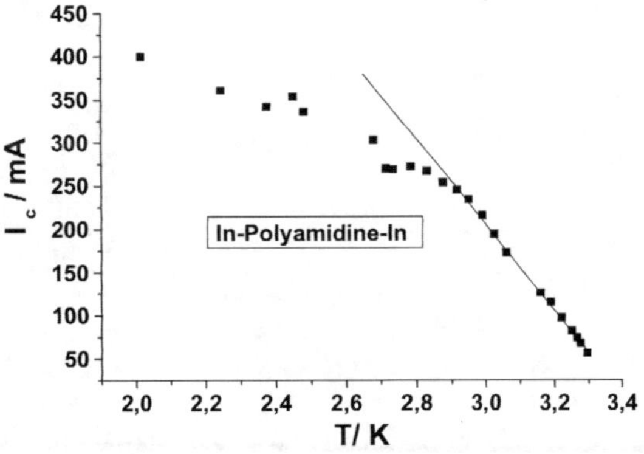

FIGURE 5. The dependence I_c against T for S-P-S construction with high capacitance.

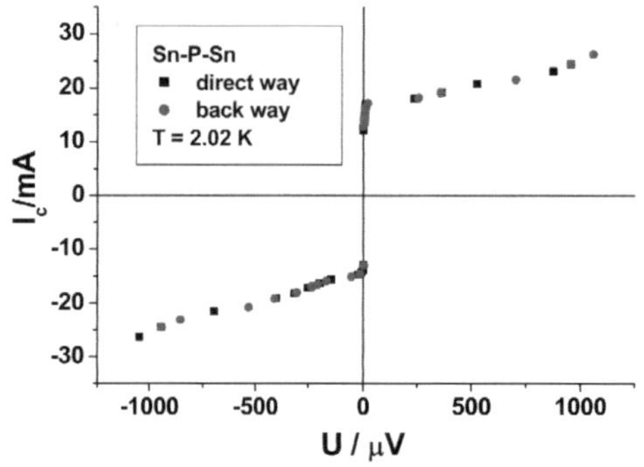

FIGURE 6. CVC of a S-P-S structure with low capacitance.

The CVCs for a hard polymer – polyimide at about the same external pressure with corresponding I_c(T) dependence are shown in Fig 6. In this case CVC has no the hysterisis loop that points to a small value of capacitance. The I_c(T) dependence was also transformed and became similar to *T*-dependence of point contact between two superconductors (Fig.7) in agreements with [12].

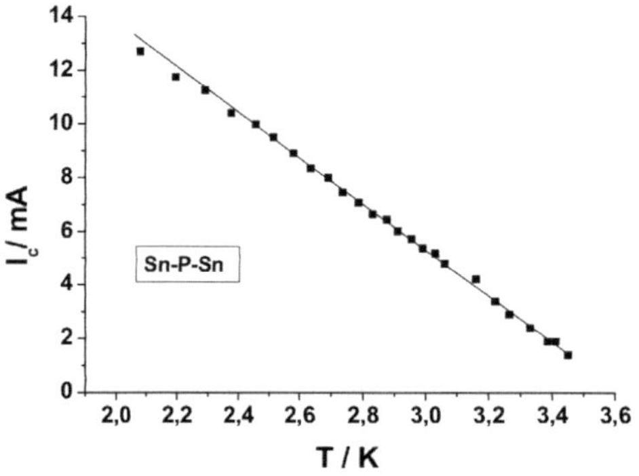

FIGURE 7. The dependence I_c against *T* for a S-P-S sandwich with low capacitance.

Figure 8 presents I_c(T) dependence for the same polymer under different pressure. With increasing pressure the I_c(T) dependence got linear. This means that polymer capacitance decreases due to the decreasing distance between the localized states.

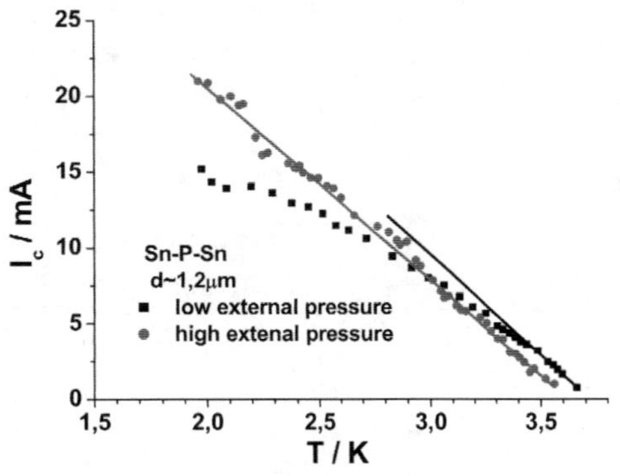

FIGURE 8. Dependence I_c versus T for the same S-P-S structure under different pressures.

Figure 9 shows how the S-P-S junction with resonant tunneling responds to microwave radiation [13]. Sharp Shapiro steps with accuracy of 100 nV are seen at the voltage given by $V_n = n\frac{h}{2e}f$ for n = 1,2,..6., here f is frequency. This behavior is characteristically for a single Josephson junction. Current jumps are very sharp as observed in a homogeneous system. It must be emphasized that the existence of the Josephson effect and Shapiro steps in any junction are a direct proof for superconducting properties of the system.

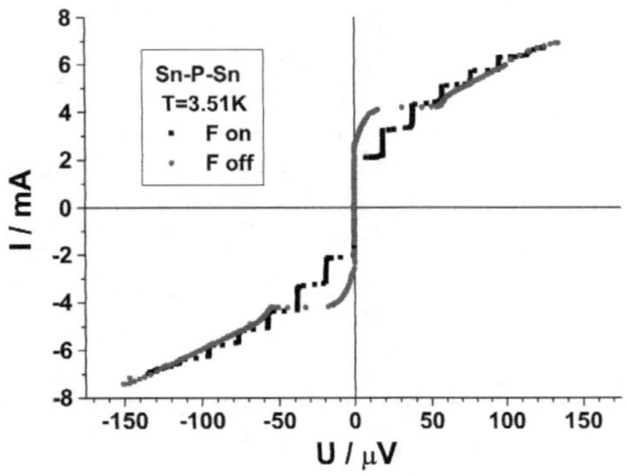

FIGURE 9. Shapiro steps induced by microwave (f = 9.3 GHz) radiation (black points). The steps occur exactly at the voltage given by $V_n = n\frac{h}{2e}f$ and correspond to a single Josephson junction.

A SPS junction is a very complex object because it consists a Superconductor-(Insulator-Macromolecul-Insulator)$_n$-Superconductor stack with a large number of n. Analysis of our experimental data in the frame of resonant tunneling model gave a satisfactory explanation of the dependence of I_c on ξ_s. We used the simple theoretical proximity model of superconductivity [10]. However, our results do not reject any existing intrinsic superconductivity in polymers [14]. But for this observation it is necessary to find out the preparation conditions of junctions on the base of polymers.

Acknowledgments

We wish to thank Dr. L.N. Ionov for assistance in microwave radiation experiments and Dr. D.Eagles and Prof. M.Yu. Kupriyanov for discussion.

References

1. J. Lowell and A. C. Rose-Innes, *Adv. Phys*, **29**, 947-1023 (1980).
2. C. B. Duke, T. I. Fabish, *Phys. Rev. Let*ters **37**, 1075-1078 (1976).
3. N. S. Enikolopyan, L. N. Grigorov, S. G. Smirnova, *JETP Letters* **49**, 371- (1989).
4. A. N. Ionov, R. Rentzsch, M. N. Nikolaeva *Phys. Stat. Sol. (C)* **5**, 730-734 (2008).
5. R. Rentzsch and A. N. Ionov *Phys. Stat. Sol. (C)* **5**, 735-739 (2008).
6. C. B. Duke, U. R. Salaneck, T. I. Fabish, J. J. Ritsko, H. R. Thomas and A. Paton, *Phys Rev. B* ***18***, 5717–5739 (1978).
7. V. A. Zakrevskii, A. N. Ionov and A. N. Lachinov, *Tech. Phys. Letters* 24, 539 - (1998).
8. A. N. Ionov, M. S. Dunaevskii, M. N. Nikolaeva, and R. Rentzsch *Annalen der Physik* **18**, 959-962 (2009).
9. L. G. Aslamazov and M. V. Fistul' *Sov. Phys. LETP* **56**, 666- (1982).
10. P. G. De Gennes, *Rev. Mod. Phys.* **36**, 225- (1964)
11. K. K. Likharev *Dynamics of Josephson Junctions and Circuits*, New York, Gordon and Breach, 1986, pp. .
12. A. A. Golubov, M. Yu. Kupriyanov, E. Il'ichev *Rev. Mod. Phys.*, **76**, 411-469 (2004).
    M. P. Brown and K. Austin, *Appl. Phys. Letters* **85**, 2503-2504 (2004).
13. S. Shapiro, *Phys. Rev. Letters,* **11**, 80 - (1963).
14. D. Eagles,

Journal of Physics: Conference Series **376** (2012) 012024 doi:10.1088/1742-6596/376/1/012024

Interference quantum correction to conductivity of $Al_xGa_{1-x}As/GaAs$ double quantum well heterostructures near the balance

A. V. Germanenko[1], **G. M. Minkov**[1,2], **O. E. Rut**[1] and **A. A. Sherstobitov**[1,2]

[1] Institute of Natural Sciences, Ural Federal University, Ekaterinburg 620000, Russia
[2] Institute of Metal Physics RAS, Ekaterinburg 620990, Russia

E-mail: `alexander.germanenko@usu.ru`

Abstract. We present the results of experimental investigations of the interference quantum correction to the conductivity of the gated double quantum well $Al_xGa_{1-x}As/GaAs/Al_xGa_{1-x}As$ heterostructures. Analyzing the positive magnetoconductivity we obtain the interwell transition rate and the phase relaxation rate under the conditions when one and two quantum wells are occupied. It has been found that the interwell transition rate resonantly depends on the difference between the electron densities in the wells in accordance with the theoretical estimate. The central point, however, is that the dephasing rate in the lower quantum well is independent of whether the upper quantum well contributes to the conductivity or not. The results obtained are interpreted within framework of the recent theory for the dephasing and electron-electron interaction in the double well structures [Burmistrov I S, Gornyi I V and Tikhonov K S 2011 *Phys. Rev.* B **84** 075338].

1. Introduction

The quantum corrections to the conductivity of the two-dimensional (2D) electron gas with the single-valley parabolic energy spectrum are extensively investigated last three decades. A reasonably good agreement between experimental and theoretical results evident at relatively high conductivity, $\sigma \gtrsim (10-15)\,G_0$, where $G_0 = e^2/\pi\, h$, attests adequate understanding of the role of the quantum interference and electron-electron (e-e) interaction in the transport properties of 2D systems. The quantum corrections in the double quantum well (DQW) structures are studied significantly less. Specific features of the corrections in this case are governed by relationship between the following parameters: the transport time τ, the phase relaxation time τ_ϕ, and the interwell transition time τ_{12}, the temperature length $L_T = \sqrt{D/T}$ (where D is the diffusion coefficient), the interwell distance d, and the screening length r_0. In the limiting cases such structures can look like a single quantum well structure (at $\tau_{12} \ll \tau_\phi$, $1/T$) or a structure with two uncoupled wells (at $\tau_{12} \gg \tau_\phi$, $1/T$ and $d \gg r_0$). The intermediate cases are more interesting and diversified. Concerning the interaction correction, the gain of the interaction in a multiplet channel can be so significant that it can lead to the change of the sign of the temperature dependence of the conductivity from dielectric to metalliclike as it takes place in the 2D systems with double valley energy spectrum. The interference correction, even without interwell transitions ($\tau_{12} \gg \tau_\phi$), can differ from the case of non-interacting wells due to inelasticity of the e-e interaction of carriers in the different wells. This interaction can change the dephasing rate significantly.

2. Experimental details

Two structures grown by the molecular beam epitaxy on a semiinsulating GaAs substrate were investigated. The first structure 3243 consists of 200 nm-thick undoped GaAs buffer layer, 50 nm undoped $Al_{0.3}Ga_{0.7}As$ layer, two 8 nm GaAs quantum wells separated by 10 nm $Al_{0.3}Ga_{0.7}As$ barrier, 70 nm $Al_{0.3}Ga_{0.7}As$ undoped layer. The thickness of the GaAs cap layer is 130 nm. The main doping δ layer of Si is situated in the barrier center. Because the gate voltage in such a type of structures can effectively decrease the electron density only, an additional Si δ layer is located above the upper quantum well at distance of 18 nm from the well interface. The second structure 3154 differs by the doping level and has lower electron density and mobility. Such structure design allows us to have the close mobilities in the wells at equal electron density that is very important for reliable determination of the phase relaxation time as will be discussed below. The samples were mesa etched into standard Hall bars and then an Al gate electrode was deposited by thermal evaporation onto the cap layer through a mask.

3. Interference quantum correction

It is well known that the suppression of the weak localization (WL) by in the transverse magnetic field $B \parallel n$, where n is the normal to the structure plane results in negative magnetoresistivity. In single quantum well (SQW) systems the shape of magnetoconductivity $\Delta\sigma(B) = 1/\rho_{xx}(B) - 1/\rho_{xx}(0)$ is described by the following expression [1, 2, 3]:

$$\frac{\Delta\sigma(B)}{G_0} = \alpha\mathcal{H}\left(\frac{\tau_\phi}{\tau}\frac{B}{B_{tr}}\right), \quad \mathcal{H}(x) = \psi\left(\frac{1}{2}+\frac{1}{x}\right) + \ln x, \tag{1}$$

where $G_0 = e^2/\pi h$, $B_{tr} = \hbar/4eD\tau$ is a transport magnetic field, $\psi(x)$ is a digamma function, α is prefactor, which appears due to not rigorous fulfilment of the diffusion approximation $\tau/\tau_\phi \ll 1$ [4], due to the magnetic field dependence of the dephasing time [5] and the contributions of the second order corrections [6]. As for the DQW systems, the shape of the magnetoconductivity curve is widely varied. For instance, even in the case of low interwell transition rate, when $\Delta\sigma(B)$ is barely the sum of the contributions from each well

$$\frac{\Delta\sigma(B)}{G_0} = \alpha_1\mathcal{H}\left(\frac{\tau_{\phi1}}{\tau_1}\frac{B}{B_{tr1}}\right) + \alpha_2\mathcal{H}\left(\frac{\tau_{\phi2}}{\tau_2}\frac{B}{B_{tr2}}\right), \tag{2}$$

it depends, as seen, on the large number of parameters: the two phase relaxation times, $\tau_{\phi1}$ and $\tau_{\phi2}$, the two transport times, τ_1 and τ_2, and two transport magnetic fields, B_{tr1} and B_{tr2}. Obviously it is impossible to find six parameters while fitting the single smooth curve. Reliable obtaining of the phase relaxation time is possible only when the corresponding parameters τ_i, $\tau_{\phi i}$, and $B_{tr\,i}$ ($i = 1, 2$) in the wells are close to each other.

In order to find experimentally the gate voltage, at which the DQW structure is in the balance (i.e., electron densities in the wells n_1 and n_2 are identical) and to ascertain that the mobility of electrons in the wells are close to each other, we have analyzed the magnetic field dependences of the Hall coefficient, the classical magnetoresistivity and the Shubnikov-de Haas effect (for more details, see [7]). The parameters obtained are listed in table 1. From here on the regime when the upper quantum well is depleted and only the lower well is occupied is referred to as SQW regime.

The magnetic field dependences of $\Delta\sigma(B)$ at $T = 1.35$ K for different gate voltages V_g are presented in Fig. 1(a). At first sight all the curves are the same in the shape and no peculiarity

Table 1. Parameters of the structures investigated.

Structure	#3243		#3154	
Regime	SQW	balance	SQW	balance
V_g(V)	−4.1	−1.5	−3.6	−2.0
n (10^{11} cm^{-2})	7.0	7.5	4.0	4.5
μ (10^3 cm^2/Vs)	14.5	15.0	4.8	6.5
$\tilde{\gamma}_{t,0}$	0.30	—	0.35	—
$\tilde{\gamma}_s$	—	−0.77	—	−0.77
$\tilde{\gamma}_t$	—	0.30	—	0.35
$\tilde{\gamma}_v$	—	0.007	—	0.009

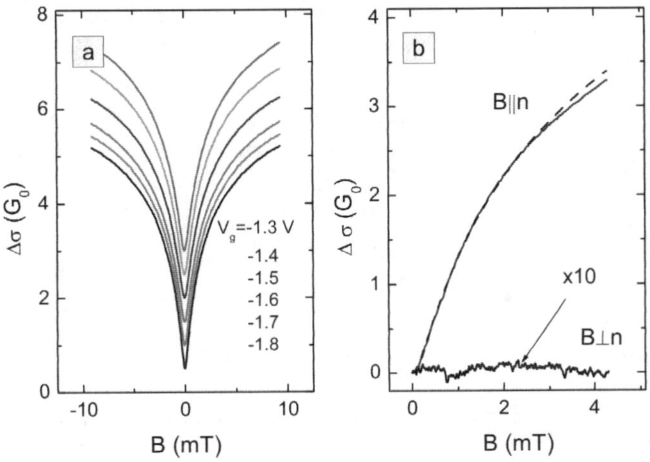

Figure 1. (a) The interference-induced positive magnetoconductivity for different gate voltages near the balance. For clarity, the curves are shifted in the vertical direction. (b) The magnetoconductivity measured at $V_g = -1.5$ V for two orientations of the magnetic field. Solid lines are the data, the dashed line is the best fit by (2) with the parameters $\alpha = 1.9$, $\tau_\phi = 85$. $T = 1.35$ K for all the cases, $B_{tr} = 8.4$ mT.

is observed in the balance. This, indeed, should be happening when the interwell transition rate is low, $1/\tau_{12} \ll 1/\tau_\phi$, and $\Delta\sigma(B)$ is the sum of the contributions from each well.

Closeness of the mobilities in the balance means closeness of transport fields, $B_{tr1} = B_{tr2} = B_{tr}$ so that $\Delta\sigma(B)$ should be described by (1), which in this case reduces to expression (1) with $\alpha = 2$. Therefore, let us start analysis from $V_g = -1.5$ V. The result of the best fit within the magnetic field range $(0 - 0.3)B_{tr}$ for $T = 1.35$ K is shown in Fig. 1(b). It is evident that (1) perfectly fits the data with the prefactor which value is really about two. The temperature dependences of the fitting parameters τ_ϕ and α are presented in Fig. 2. It is seen that the T dependence of τ_ϕ is close to $1/T$ (some deviation will be discussed below). Slight decrease of the prefactor with the growing temperature is a consequence of the lack of the diffusion regime due to the decrease of the τ_ϕ to τ ratio [4] from 85 at $T = 1.35$ K to 30 at $T = 4.2$ K.

Let us now analyze the data in the vicinity of $V_g = -1.5$ V. The gate voltage dependence of the fitting parameter $\tau_\phi^\star$ presented in Fig. 3(a) by the diamonds exhibits the sharp minimum at $V_g = -1.5$ V (the superscript * implies that the τ_ϕ value found from the fit can differ from the true dephasing time). It stands to reason that this minimum results from the interwell transitions neglected in Eq. (2). Theoretically, their role in the positive magnetoconductivity was considered in Refs. [8] and [9]. However, our attempts to use the expressions from these papers for the data fit have failed. The values of τ_ϕ and τ_{12} are obtained with very large uncertainty. This is

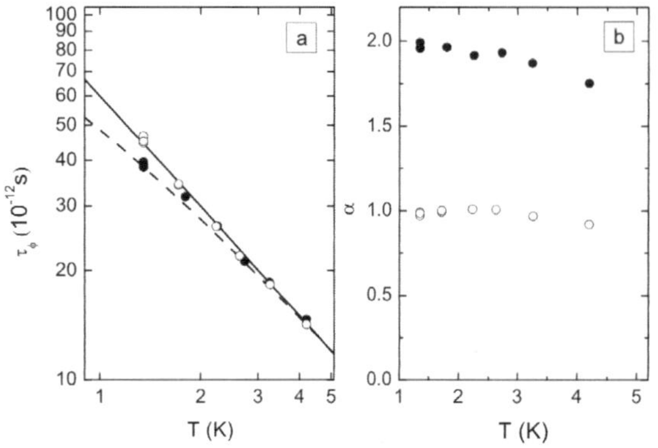

Figure 2. The temperature dependences of τ_ϕ (a) and α (b) found from the fit of the magnetoconductivity curves by (1). Solid symbols are obtained at $V_g = -1.5\ V$ when the structure is in the balance. Open symbols correspond to $V_g = -3.7\ V$ when the conductivity over the upper well is negligible. The solid and dashed lines are the functions $6 \times 10^{-11}/T$ and $(T/6.4 \times 10^{-11} + 1/2 \times 10^{-10})^{-1}$, respectively.

because the shape of the magnetoconductivity curve in actual case is mainly controlled by some combination of τ_ϕ and τ_{12} but not by τ_ϕ and τ_{12} individually. It becomes clear when considering the parabolic (at $B \to 0$) and logarithmic (at $B \gg B_{tr}\tau/\tau_\phi$) asymptotics of Eq. (24) in Ref. [9] at $\tau_{12} > 2\,\tau_\phi$. They coincide with corresponding asymptotics, which one can extract from Eq. (1) by using $\alpha = 2$ and $\tau_\phi^\star = (1/\tau_\phi + 1/\tau_{12})^{-1}$ instead of τ_ϕ. So, the sum $1/\tau_\phi + 1/\tau_{12}$, but not the rates $1/\tau_\phi$ and $1/\tau_{12}$ separately are experimentally obtained when the interwell transitions are relatively rare.

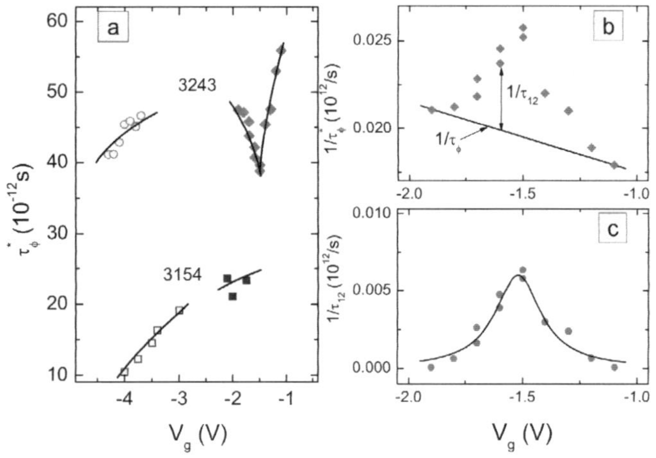

Figure 3. (a) The values of the fitting parameter $\tau_\phi^\star$ plotted against the gate voltage for structures 3243 and 3154. Solid symbols are obtained near the balance. The open symbols correspond to SQW regime. The lines are provided as a guide to the eye. (b) The $1/\tau_\phi^\star$ vs V_g plot for structure 3243 near the balance. (c) The V_g dependence of the transition rate for structure 3243. Symbols are the experimental data, the line is equation (3).

According to Ref. [9], the interwell transition rate resonantly depends on the difference between the Fermi energies in the wells. When the electron density in one well (in lower one in our case) depends on the gate voltage weakly, the V_g dependence of the transition rate is given by the following expression

$$\frac{1}{\tau_{12}(V_g)} = \frac{1}{\tau_{12}^b}\left[1 + \left(\frac{\pi\hbar}{m}\frac{dn}{dV_g}(V_g - V_g^b)\tau\right)^2\right]^{-1}, \tag{3}$$

which can be directly compared with the data. Here, $1/\tau_{12}^b$ is the rate in the balance and m is the electron effective mass. Experimentally, the rate $1/\tau_{12}$ near the balance can be found as the difference between $1/\tau_\phi^\star$ and $1/\tau_\phi$, if one supposes that the interwell transitions do not contribute to $\tau_\phi^\star$ at $V_g = -1.8$ V and $V_g = -1.2$ V, and $1/\tau_\phi$ linearly depends on V_g between these gate voltages as shown in Fig. 3(b). The results are presented in Fig. 3(c) by the circles. As seen the experimental V_g dependence of $1/\tau_{12}$ is really close to the theoretical one calculated from Eq. (3) with $\tau_{12}^b = 1.7 \times 10^{-10}$ s and $\tau = 5.5 \times 10^{-13}$ s found from the mobility value at $V = V_g^b$ given in table 1.

Thus, we obtain from the above analysis that the dephasing time τ_ϕ in the balance is about 5×10^{-11} s at $T = 1.35$ K, whereas the interwell transition time is approximately three times larger, $\tau_{12} \simeq 1.7 \times 10^{-10}$ s. The fact that $\tau_\phi^\star$ is the combination of τ_ϕ inversely dependent on the temperature and τ_{12}, which is independent of the temperature, explains the tendency to saturation of the $\tau_\phi^\star$ vs T dependence at low temperature [see the solid symbols and the dashed curve in Fig. 2(a)].

It is impossible to find the values of $\tau_{\phi 1}$ and $\tau_{\phi 2}$ within the gate voltages range from -3.5 V to -2.0 V because six independent parameters govern the magnetoconductivity [see discussion below Eq.(2)]. However, it can be easily done at lower gate voltage, $V_g < -3.7$ V, when the conductivity and, thus, the magnetoconductivity are determined by the lower well only. The $\Delta\sigma$ vs B curve for this case is well fitted by Eq. (1) with the prefactor α, which value is close to 1 [see Fig. 2(b)]. Therewith the temperature dependence of τ_ϕ, as seen from Fig. 2(a), is close to $1/T$.

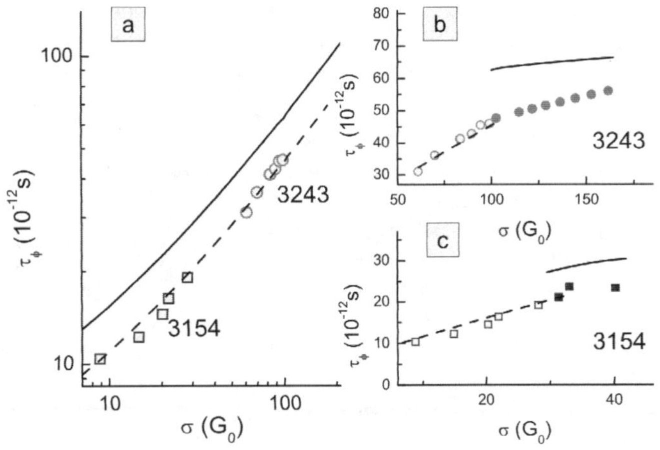

Figure 4. (a) The conductivity dependence of the phase relaxation time for the SQW regime. The solid line is calculated according [10]. The dashed line is obtained by multiplying the solid curve by the factor 0.7. (b) and (c) The phase relaxation time plotted against the conductivity per the well for the SQW regime (open symbols, dashed lines) and for the DQW regime near balance (solid symbols, solid lines). The dashed lines are calculated in the same way as in the panel (a). The solid lines are the theory for the DQW regime.

As known, the dephasing time τ_ϕ should mainly depend on the conductivity, at least for the SQW structure [10]. In Fig. 4(a), we show the data obtained for both structures in the SQW regime. In the same figure, the theoretical dependence [10] is depicted by the solid line. As seen, being multiplied by the factor 0.7 it perfectly describes the data within the entire conductivity range. We believe that the appearance of the additional factor can result from the fact that the expression used for the data treatment does not take into account the magnetic field dependence of τ_ϕ, the contributions of the second order corrections, and the finiteness of the τ_ϕ to τ ratio.

Let us again direct the reader's attention to the data obtained near the balance. These data are presented in Figs. 4(b) and 4(c) by the solid symbols. As seen the experimental τ_ϕ values in SQW and DQW regimes fall on the common curve. This means that the inelastic interaction of an electron in the one well with electrons in the other one does not contribute noticeably to the dephasing rate. At first glance the last seems very strange because the distance between the wells is about screening length so that the interaction between the carriers located

Journal of Physics: Conference Series **376** (2012) 012024 doi:10.1088/1742-6596/376/1/012024

in the different wells should not be significantly less than that for carriers in the one well. On the other hand, the inelastic interaction of an electron with the electrons in the other well is not sole effect influencing the dephasing rate. Another effect is an additional screening of the interaction between the electrons in the one well by the carriers located in the other one. The former effect should lead to the τ_ϕ decrease, while the second one to its increase. In our previous paper [7] we have compared the data obtained with the results of simple theory which takes into account both these effects in singlet channel only. Remarkable discrepancy has been mentioned. The theory predicts the increase in the dephasing time in double-layer structures as compared with the single-layer case [shown by solid lines in figures 4(b) and 4(c)], which is not observed experimentally.

In recent paper [11], the theory, which takes into account contributions to the dephasing rate from the interaction both in singlet and multiplet channels has been developed. According to [11] the dephasing rate at the balance can be found from equation

$$\frac{\tau_\phi^0}{\tau_\phi} = \frac{\mathcal{A}}{\mathcal{A}_0}\frac{\ln T\tau_\phi}{\ln T\tau_\phi^0},\ \mathcal{A}_0 = 1 + \frac{3\tilde{\gamma}_{t,0}^2}{2+\tilde{\gamma}_{t,0}},\ \mathcal{A} = \frac{1}{2}\left(1 + \frac{\tilde{\gamma}_s^2}{2+\tilde{\gamma}_s} + 6\frac{\tilde{\gamma}_t^2}{2+\tilde{\gamma}_t} + 8\frac{\tilde{\gamma}_v^2}{2+\tilde{\gamma}_v}\right), \tag{4}$$

where τ_ϕ^0 denotes the dephasing time of a single layer, and $\tilde{\gamma}_{t,0}$, $\tilde{\gamma}_s$, and $\tilde{\gamma}_v$ are the interaction constants. Using the values of interaction constants listed in table 1 for our samples, we obtain the value 0.77 for $\mathcal{A}$ to $\mathcal{A}_0$ ration instead of 0.75 in [7] for both samples. The difference between these values is very small and does not practically change the run of sold curves in Fig. 4(c). Thus, the taking into account of the interaction contribution in the multiplet channel dos not explain the experimental results concerning the dephasing rate in the DQW heterostructures.

4. Conclusion

We have studied the interference quantum correction in the gated double quantum well $Al_xGa_{1-x}As/GaAs/Al_xGa_{1-x}As$ heterostructures. Analyzing the positive magnetoconductivity we have obtained the interwell transition rate and the phase relaxation rate under the conditions when one and two quantum wells are occupied. It has been found that the dephasing rate in the lower quantum well is independent of whether the upper quantum well contributes to the conductivity or not. This observation is inconsistent with theory, which predicts the increase of the dephasing time in double layer structures as compared with the single layer case.

Acknowledgments

We would like to thank I. S. Burmistrov and I. V. Gornyi for illuminating discussions. This work has been supported in part by the RFBR (Grants No. 08-02-00662, No. 09-02-00789, and No. 10-02-00481, and No. 10-02-91336).

References

[1] Hikami S, Larkin A I and Nagaoka Y 1980 *Prog. Theor. Phys.* **63** 707 – 710
[2] Chakravarty S and Schmid A 1986 *Phys. Reports* **140** 193 – 236
[3] Wittmann H P and Schmid A 1987 *J. Low Temp. Phys.* **69** 131 – 149
[4] Minkov G M, Germanenko A V, Larionova V A, Negashev S A and Gornyi I V 2000 *Phys. Rev. B* **61** 13164 – 13171
[5] Germanenko A V, Larionova V A, Gornyi I V and Minkov G M 2007 *Int. J. Nanosci.* **6** 261264
[6] Minkov G M, Germanenko A V and Gornyi I V 2004 *Phys. Rev. B* **70** 245423
[7] Minkov G M, Germanenko A V, Rut O E, Sherstobitov A A, Bakarov A K and Dmitriev D V 2010 *Phys. Rev. B* **82** 165325
[8] Averkiev N S, Golub L E and Pikus G E 1998 *Fiz. Tekh. Poluprovodn.* **32** 1219
[9] Raichev O E and Vasilopoulos P 2000 *J. Phys.: Condens. Matter* **12** 589–600
[10] Narozhny B N, Zala G and Aleiner I L 2002 *Phys. Rev. B* **65** 180202
[11] Burmistrov I S, Gornyi I V and Tikhonov K S 2011 *Phys. Rev. B* **84** 075338

Journal of Physics: Conference Series **376** (2012) 012025 doi:10.1088/1742-6596/376/1/012025

Quantum Coherence in Photosynthetic Exciton Dynamics

Peter Nalbach, Michael Thorwart

I. Institut für Theoretische Physik, Universität Hamburg, Jungiusstraße 9, 20355 Hamburg, Germany

E-mail: `Peter.Nalbach@physik.uni-hamburg.de`

Abstract.
We present numerically exact results for the electronic energy transfer dynamics in the Fenna-Matthews-Olson (FMO) molecular aggregate. In particular, we determine its single excitation subspace dynamics within an open quantum dynamics approach which is based on parameters for the dipolar couplings, site energies and their respective fluctuations. We study and compare the dynamics for parameters according to several models available in the literature. Depending on the model used we find quantum coherent dynamics with coherence times exceeding experimentally observed times but also shorter times for other models. Even strongly overdamped decay of the site populations with no coherent population dynamics is observed for model parameters based on atomistic molecular dynamics. Thus, theoretical understanding at the present stage can neither support nor rule out a picture of quantum coherent energy transfer in natural photosynthesis.

1. Introduction

Recently, experiments have been reported in which long-lived coherence effects have been observed in the ultrafast exciton dynamics in photosynthetic biomolecules. These results caused tremendeous renewed interest in the longstanding question whether quantum coherence effects have any functional significance in natural biological systems under physiological conditions. Photosynthesis [1] starts with the absorption of a photon by a pigment and the formation of an exciton, followed by its transfer to the reaction center, where charge separation takes place. The excitation transfer has traditionally been regarded as an incoherent hopping between molecular sites [2].

A few years ago, Engel et al. [3, 4] have reported long-lasting beating signals in time-resolved optical two-dimensional spectra of the Fenna-Matthews-Olson (FMO) [5, 6] complex. These have been interpreted as evidence for quantum coherent energy transfer via delocalized exciton states. FMO's function is to transfer energy from the chlorosome, a large vesicle of pigments, to the reaction center. FMO consists of three identical subunits, each with eight bacteriochlorophyll molecular sites. Quantum coherence times of more than 660 fs at 77 K [3] and about 300 fs at physiological temperatures [4] have been reported. Similar experimental findings were observed in marine cryptophyte algae [7].

The question how quantum coherence can survive over such long times in these biomolecular environments with strong vibronic modes and a surrounding strongly coupled polar solvent is open. Full quantum theoretical modeling of the real-time dynamics is beyond current computer powers due to the large cluster size and strong non-Markovian fluctuations. Studies of the excitation dynamics, therefore, rely on simplified models [8] of few chromophore sites which interact by dipolar couplings and which are exposed to environmental fluctuations of the bacteriochlorophyll molecule, the surrounding protein and the solvent [1, 2].

From simple donor-acceptor toy models one realizes that standard Redfield-type or other weak coupling approaches fail for typical environmental fluctuation spectra [9, 10]. Nevertheless, weak coupling approaches including various extensions have been employed to investigate the question whether quantum coherence is functionally relevant for the near unity quantum efficiency of energy transfer [11, 12, 13]. In order to study the origin of the long lived coherence, methods which can describe the non-Markovian behavior exactly have to be employed [14]. Again, donor-acceptor toy model studies showed that either temporal correlations of environmental fluctuations, with time scales exceeding the internal electronic coupling time scale, or spatial correlations which extend between the chromophores, which exchange energy, can cause long lived coherence [15, 16, 17, 18].

In order to study excitonic energy transfer in FMO, model descriptions including an excitonic Hamiltonian and environmental fluctuation spectra are necessary. Based on an analysis of optical experimental data Adolphs and Renger determined site energies, dipolar couplings between pigments and the fluctuation spectrum for the site energies for a subunit of the FMO trimer [19]. The latter includes, besides a broad continuous spectrum resulting from the solvent fluctuations, additionally a single low-frequency vibronic mode. This spectrum goes back to experimental measurements of Wendling et al. [20] who have measured fluorescence line-narrowing at low temperature. Adolphs and Renger, however, only considered the discrete vibronic state with the lowest frequency (besides the continuous background). This Adolphs / Renger (AR) model was further simplified by Cho et al. [21] (CF-model) and also by Ishizaki and Fleming [22] (IF-model) by considering a phenomenological environmental fluctuation spectrum.

Recently, Olbrich et al. have used extensive classical molecular dynamics simulations for the FMO aggregate including the environmental solvent to determine site energies, electronic couplings and their respective fluctuational characteristics [23] (OK-model). In a follow-up paper, Olbrich et al. [24] have analyzed the properties of the fluctuations in more detail and have determined explicitly the environmental fluctuation spectra necessary for the full FMO dynamics. Noteworthy, they obtain different spectra for each site. Each spectrum consists of two continuous solvent contributions and a set of up to 13 vibronic modes. Thus, the OK-model for FMO includes much more details than the IF-, the CF- or the AR-model but is solely obtained from classical MD simulations.

Ishizaki and Fleming used a second-order cumulant time-nonlocal quantum master equation to determine the site population dynamics in the single excitation subspace [22] and found quantum coherent dynamics with coherence times as observed experimentally for the FMO complex. The present authors used the quasi adiabatic path integral approach (QUAPI) to study the dynamics of the AR model [25] where we found coherence times roughly by a factor of two smaller than observed in the FMO complex.

In the present report, we use the OK-model for the FMO and determine the population dynamics numerically exactly using QUAPI for a monomeric subunit of the trimer. We find that the time evolution of the site populations is overdamped at room temperature as well as at 77 Kelvin. Furthermore we study the dynamics in the CF-model where we observe coherence times exceeding experimentally observed ones.

The present work is organized as follows: In the following section we briefly summarize for completeness the various models for FMO and discuss the resulting site energies, coupling strengths and environmental fluctuation spectra. Then, we briefly summarize the numerical method used to solve the dynamics of the FMO model coupled to environmental fluctuations. We then present results for a monomeric subunit of the FMO trimer for each of the introduced

Journal of Physics: Conference Series **376** (2012) 012025 doi:10.1088/1742-6596/376/1/012025

models. A comparison of the results from the various models shows that we are not able to conclude on an answer to the question whether the excitation transfer is quantum mechanically coherent on time scales up to 500 fs at physiological temperatures.

2. Model

2.1. Model - Hamiltonian

The FMO complex is a trimer consisting of identical, weakly interacting monomers, each containing eight bacteriochlorophyll$a$ (BChl$a$) molecular sites. Excitation energy is transfered between these pigments. They are embedded in a large protein complex. Typically, each of them is reduced to its two lowest electronic levels. Thereby, their excited states are electronically coupled. Recombination is negligibly slow ($\sim$ ns) compared to exciton transfer times ($\sim$ ps) and multiple excitation is unlikely under natural conditions. Thus, the excitation dynamics is reliably described within the 1-exciton subspace.

Adolphs and Renger [19] have determined the site energies and dipolar couplings leading to the Hamiltonian

$$H_{\text{FMO-AR}} = \begin{pmatrix} 240 & -87.7 & 5.5 & -5.9 & 6.7 & -13.7 & -9.9 \\ & 315 & 30.8 & 8.2 & 0.7 & 11.8 & 4.3 \\ & & 0 & -53.5 & -2.2 & -9.6 & 6.0 \\ & & & 130 & -70.7 & -17.0 & -63.3 \\ & & & & 285 & 81.1 & -1.3 \\ & & & & & 435 & 39.7 \\ & & & & & & 245 \end{pmatrix} \tag{1}$$

in units of cm^{-1} in site representation for an FMO monomer of *C. tepidum*. We define the lowest site energy as reference. Adolphs and Renger give also parameters for a monomeric subunit of the FMO trimer which are rather close to their monomer parameter set (therefore we omit this parameter set in the following discussion completely). Note that the description by Adolphs and Renger is based on seven pigments for each monomeric subunit of the FMO trimer. The excistence of the eighth pigment was only realized recently [26].The eighth pigment has, however, little influence on coherence which is dominated by the strongly coupled pigments 1 and 2.

Olbrich et al. obtain for a monomeric subunit of the trimer

$$H_{\text{FMO-OK}} = \begin{pmatrix} 185.4 & -80.3 & 3.5 & -4.0 & 4.5 & -10.2 & -4.9 & 21.0 \\ & 80.6 & 23.5 & 6.7 & 0.5 & 7.5 & 1.5 & 3.3 \\ & & 0.0 & -49.8 & -1.5 & -6.5 & 1.2 & 0.7 \\ & & & 112.8 & -63.4 & -13.3 & -42.2 & 1.2 \\ & & & & 64.5 & 55.8 & 4.7 & 2.8 \\ & & & & & 88.6 & 33.0 & -7.3 \\ & & & & & & 491.7 & -8.7 \\ & & & & & & & 217.6 \end{pmatrix} \tag{2}$$

in units of cm^{-1} in site representation. Here, we actually consider the seven chromophores of one monomeric subunit and the eighth chromophore of the one neighboring subunit for which the distance $1-8$ is minimal. This eighth pigment is rather strongly coupled to the monomeric subunit under consideration whereas the actual eighth chromophore of this subunit is only weakly coupled to it.

In both model Hamiltonians, site 3 has lowest site energy which motivates the fact that site 3 is generally believed to be 'exit' site from which the energy is transfered from the FMO complex into the reaction center. There are, however, two noteworthy differences. Firstly, the AR-Hamiltonian (1) has sites 1 and 6 with highest energy whereas the OK-Hamiltonian (2) has

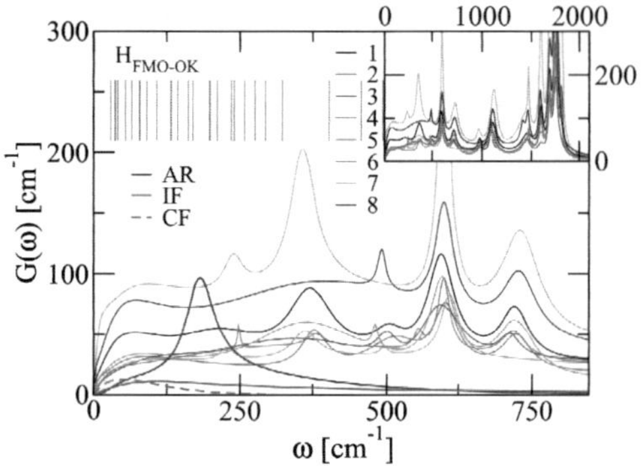

Figure 1. Environmental spectral densities according to Olbrich et al. for the FMO trimer are plotted by a full line for each site j. Additionally, the spectra of CF, IF and AR are shown. The vertical blue lines in the upper left corner depict the energy differences in the Hamiltonian (2).

sites 7 and 8 with highest energy. Secondly, in the OK-Hamiltonian (2) the dipolar couplings tend to be smaller than in the AR-Hamiltonian (1). This already hints towards weaker coherent features in the OK-model.

2.2. Model - Fluctuation spectra

The vibrations of the BChl*a*, the embedding protein and the surrounding polar solvent are too complex for a pure microscopic description and are thus treated within the framework of open quantum systems. They induce thermal fluctuations described by harmonic modes [2] and lead to the total Hamiltonian

$$H = H_{\mathrm{FMO}} + \sum_{j=1}^{7} |j\rangle\langle j| \sum_{\kappa} \nu_{\kappa}^{(j)} q_{j,\kappa} + \sum_{j=1}^{7} \tfrac{1}{2} \sum_{\kappa} \left(p_{j,\kappa}^2 + \omega_{j,\kappa}^2 q_{j,\kappa}^2\right) \tag{3}$$

with momentum $p_{j,\kappa}$, displacement $q_{j,\kappa}$, frequency $\omega_{j,\kappa}$ and coupling $\nu_{\kappa}^{(j)}$ of the environmental vibrations at site j. Notice that each BChl*a* has its own fluctuating environment. The key quantity which determines the FMO coherence properties is the environmental spectral density $G_j(\omega) = \sum_{\kappa}(|\nu_{\kappa}^{(j)}|^2/2\omega_{j,\kappa})\delta(\omega - \omega_{j,\kappa})$.

Ishizaki and Fleming [22] used a single Ohmic spectrum with a Debye cut-off function for each site independently

$$G_{IF}(\omega) = \frac{2\lambda}{\pi} \frac{\omega\omega_{c,IF}}{\omega_{c,IF}^2 + \omega^2} \tag{4}$$

with reorganization energy $\lambda = 35\mathrm{cm}^{-1}$ and Debye relaxation time (or inverse cut-off frequency) $\omega_{c,IF}^{-1} = 50\mathrm{fs}$. This spectrum is depicted in Fig. 1 by the dark green full line (lower left corner). It includes only a continuous part describing the solvent fluctuations and omits all discrete vibronic modes from the BChl*a* or the embedding protein. The Debye relaxation time (or cut-off frequency) of this spectrum is experimentally only fixed within certain limits.

Cho et al. [21] used an exponentially cut-off Ohmic spectrum

$$G_{CF}(\omega) = \frac{\lambda}{\omega_{c,CF}} \omega e^{-\omega/\omega_{c,CF}} \tag{5}$$

with $\lambda = 35\mathrm{cm}^{-1}$ but $\omega_{c,CF} = 50\mathrm{cm}^{-1}$ ($\omega_{c,CF}^{-1} = 106\mathrm{fs}$) which is depicted in Fig. 1 by the red dashed line (lower left corner).

Wendling et al. [20] have provided fluorescence line-narrowing measurements at low temperature and have obtained the so far still most detailed experimentally determined fluctuation spectrum for FMO. Adolphs and Renger [19] give an explicit form of this spectrum,

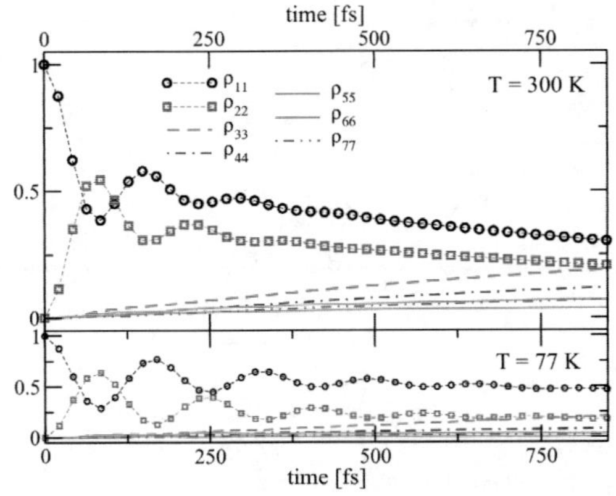

Figure 2. Time-dependent occupation probabilities for $T = 300$ K and 77 K with $\rho_{11}(0) = 1$ in the IF-model.

thereby focusing only on the discrete vibronic state with lowest frequency besides the continuous contributions, leading to

$$G_{AR}(\omega) = \omega^2 S_0 g_0(\omega) + \omega^2 S_H \delta(\omega - \omega_H) \tag{6}$$

with $S_0 = 0.5$, $S_H = 0.22$, $\omega_H = 180\text{cm}^{-1}$ and

$$g_0(\omega) = 6.105 \cdot 10^{-5} \cdot \frac{\omega^3}{\omega_1^4} e^{-\sqrt{\omega/\omega_1}} + 3.8156 \cdot 10^{-5} \cdot \frac{\omega^3}{\omega_2^4} e^{-\sqrt{\omega/\omega_2}}$$

with $\omega_1 = 0.575\text{cm}^{-1}$ and $\omega_2 = 2\text{cm}^{-1}$. This spectrum is plotted in Fig. 1 as the full violet line. There is a continuous underlying contribution plus a single overlying vibronic peak. Again independent but identical fluctuations at each site are assumed. We furthermore broadened the unphysical δ-peak and used instead a Lorentzian peak

$$\pi \omega^2 S_H \delta(\omega - \omega_H) \rightarrow \pi S_H \omega_H \frac{\omega \gamma_p}{(\omega - \omega_H)^2 + \gamma_p^2}$$

with width $\gamma_p = 29\text{cm}^{-1}$ taken as a typical value from the Olbrich et al. [24] spectra (see below).

Recently, Olbrich et al. [24] have determined by extensive molecular dynamical simulations the fluctuations of the site energies in the FMO. This allowed them to extract detailed fluctuation spectral functions $G_j(\omega)$ for each site j separately. They observe different spectra for each site. The spectra consist of two continuous contributions resulting from solvent fluctuations and up to 13 discrete vibrational modes. We plot the spectra in Fig. 1.

The spectra show discrete vibrational modes up to energies around 1800 cm^{-1} (see inset in Fig. 1). In the upper left corner (vertical blue) of Fig. 1 the energy differences between the eigenstates in the Hamiltonian (2) are indicated by vertical lines. These energy differences are (at least within a one-phonon picture) the relevant energy quanta which the FMO electronic excitations want to exchange with the environment. These energy differences do not exceed about 500 cm^{-1} and accordingly, we expect little influence of the higher energetic modes for the excitation dynamics. The same holds true for the energy differences of Hamiltonian (1). Nevertheless, some influence by the lower energetic modes could be expected. Remarkably, for the spectra at each site already the continuous solvent contribution (lowest frequency flank in the figures) is higher than for the IF-, the CF- and the AR-spectra. Thus, already from this observation, substantially shorter coherence times for the OK-spectra should be expected. The same argument would not allow to distinguish the effects of the IF-, the CF- and the AR-spectra.

Journal of Physics: Conference Series **376** (2012) 012025 doi:10.1088/1742-6596/376/1/012025

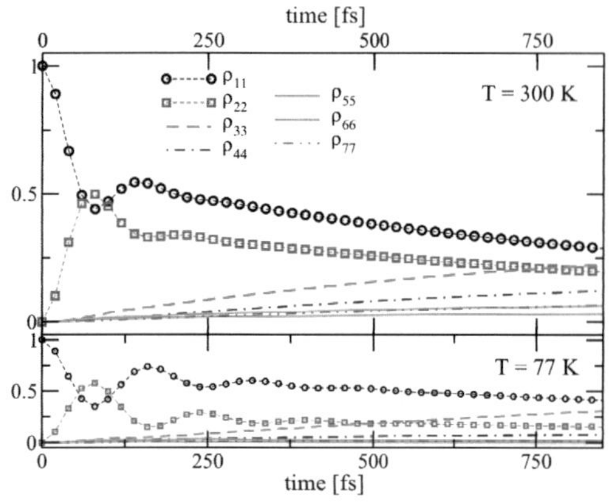

Figure 3. Time-dependent occupation probabilities for $T = 300$ K and 77 K with $\rho_{11}(0) = 1$ in the AR-model.

However, the CF-spectra has the shortest high energy tail followed by the IF-spectrum and then the AR-spectrum due to the vibrational mode. Thus, one expects coherence times longest for the CF-spectrum.

We apply the numerically exact quasi-adiabatic propagator path-integral (QUAPI) to simulate the real time exciton dynamics, thereby avoiding any approximation within the model.

3. Method

We determine the time dependent statistical operator $\rho(t)$ of the chromophoric system, consisting of either seven (CF-, IF- and AR-model) or eight (OK-model) energetic sites, using the numerically exact quasi-adiabatic propagator path-integral (QUAPI) [27, 28, 29] scheme. The QUAPI scheme was recently extended to describe quantum systems which are disturbed by multiple separate baths [17] , i.e., where each energetic site is subject to its own fluctuations and to describe various environmental initial conditions [30].

For details of this well-established iterative technique, we refer to previous works [27, 28, 29]. In brief, the algorithm is based on a symmetric Trotter splitting of the short-time propagator $\mathcal{K}(t_{k+1}, t_k)$ for the full Hamiltonian into a part depending on the system Hamiltonian and a part involving the bath and the coupling term. The short-time propagator gives the time evolution over a Trotter time slice δt. This splitting in discrete time steps is exact in the limit $\delta t \to 0$, i.e., when the discrete time evolution goes into the limit of continuous evolution. For any finite time slicing, it introduces a finite Trotter error which has to be eliminated by choosing δt small enough such that convergence is achieved. On the other side, the environmental degrees of freedom generate correlations being non-local in time. For any finite temperature, these correlations decay exponentially fast at asymptotic times, which fixes the associated memory time scale. The QUAPI scheme now defines an object called the reduced density tensor, which lives on this memory time window and establishes an iteration scheme in order to extract the time evolution of this object. All correlations are completely included over the finite memory time $\tau_{\rm mem} = K\delta t$ but are neglected for times beyond $\tau_{\rm mem}$. We have to increase accordingly the memory parameter K until convergence is found. The two strategies to achieve convergence, i.e., minimize δt but maximize $\tau_{\rm mem} = K\delta t$, are naturally counter-current, but nevertheless convergent results can be obtained in a wide range of parameters, including the cases presented in this work.

Journal of Physics: Conference Series **376** (2012) 012025 doi:10.1088/1742-6596/376/1/012025

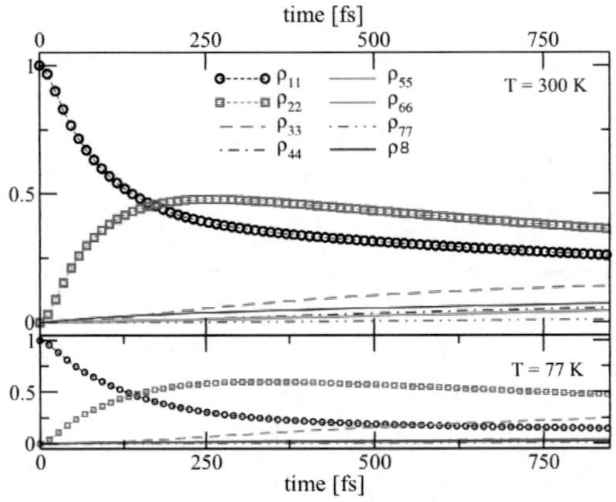

Figure 4. Time-dependent occupation probabilities for $T = 300$ K and 77 K with $\rho_{11}(0) = 1$ in the OK-model.

4. Results and Discussion

4.1. Population dynamics in the IF-model

At first, we focus on the time-dependent populations of the FMO pigment sites in the Ishizaki-Fleming model with the Hamiltonian (1) and the environmental fluctuation spectra (4) depicted in Fig. 1 by the dark green full line. We choose $T = 300$K (physiological temperature) and $T = 77$K (typical experimental temperature). Both pigments, BChl 1 and BChl 6, are oriented towards the baseplate protein and are thus believed to be initially excited (entrance sites) [31] in the natural excitation energy transfer process. However, the dipolar coupling between pigment 1 and 2 is maximal and thus the longest coherent dynamics is expected as population oscilation between these two pigments. Thus, we focus on the initial condition $\rho_{11}(0) = 1$ since we are interested in the maximal coherence times in the different FMO models.

In Fig. 2, we show the time-dependent pigment occupation probabilities $\rho_{jj}(t)$. Our results coincide with results by Ishizaki and Fleming [22]. For both temperatures, $T = 300$ K and $T = 77$ K, we observe coherent population oscilations between site 1 and 2. After 800 fs half the population is still at these two sites at room temperature but site 3, the exit site, is populated to about a 1/5. At 77 K the ocupation of the exit site is almost as large although still about half the population is in site 1.

Coherent dynamics is observed for about 400 fs at room temperature and for about 800 fs at $T = 77$ K.

4.2. Population dynamics in the AR-model

Secondly, we investigate the time-dependent populations of the FMO pigment sites in the Adolphs-Renger model with the Hamiltonian (1) and the environmental fluctuation spectra (6) depicted in Fig. 1 by the violet full line. We choose $T = 300$K (physiological temperature) and $T = 77$K (typical experimental temperature) with initial condition $\rho_{11}(0) = 1$.

In Fig. 3, we show the time-dependent pigment occupation probabilities $\rho_{jj}(t)$ (see also [25]). For both temperatures, $T = 300$ K and $T = 77$ K, we observe coherent population oscilations between site 1 and 2. As in the IF-model, after 800 fs half the population is still at these two sites at room temperature but site 3, the exit site, is populated to about 1/4 and thus slightly larger than in the IF-model. At 77 K the occupation of the exit site is almost as large but again most of the population is still at site 1.

Coherent dynamics is observed considerably shorter as compared to the IF-model. In detail, coherence is seen for about 250 fs at room temperature and for about 500 fs at $T = 77$ K.

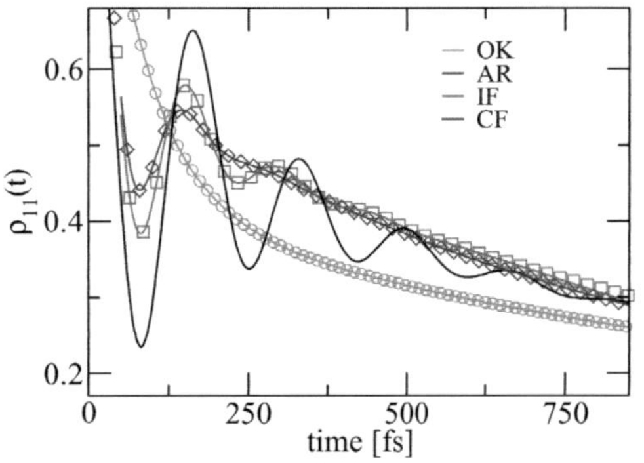

Figure 5. Time-dependent occupation probability $\rho_{11}(t)$ at $T = 300$ K with $\rho_{11}(0) = 1$ in the CF-, IF- and AR-model in comparison.

4.3. Population dynamics in the OK-model

Thirdly, we study the time-dependent populations of the FMO pigment sites in the Olbrich-Kleinekathöfer model with the Hamiltonian (2) and the environmental fluctuation spectra for each site as depicted in Fig. 1 by the full lines denoted by a number corresponding to the pigment. We choose $T = 300$K (physiological temperature) and $T = 77$K (typical experimental temperature) with initial condition $\rho_{11}(0) = 1$.

In Fig. 4, we show the time-dependent pigment occupation probabilities $\rho_{jj}(t)$. For both temperatures, $T = 300$ K and $T = 77$ K, we observe overdamped dynamics, i.e., no coherent oscillations. In contrast to both previous models more than half the population is still after 800 fs at site 1 and 2 at room temperature and the occupation of the exit site 3 is by far smaller. The population is mainly transfered from site 1 to 2 and then is stuck there. The same holds at 77 K, but surprisingly the ocupation of the exit site is larger than at room temperature and actually comparable to values from both previous models.

Coherent dynamics is not observed, neither at room temperature nor at $T = 77$ K.

4.4. Population dynamcis in the CF-model

At last we determine the time-dependent populations of the FMO pigment sites in the Cho-Fleming model with the Hamiltonian (1) and the environmental fluctuation spectra (5) depicted in Fig. 1 by the dashed red line. Due to the small cut-off frequency $\omega_{c,CF} = 50\text{cm}^{-1}$ in the CF-model convergence of the QUAPI scheme can only be achieved by explicitly extrapolating to $\delta t \to 0$ and $\tau_{mem} \to \infty$. Instead of doing so for each population at each give point in time, we use the population $\rho_{11}(t)$ and fit it using

$$f(t) = a_1 \cdot \cos(a_2 t)\exp(-a_3 t) + a_4 \cdot \exp(-a_5 t) + (1 - a_1 - a_4)$$

Seeking convergence for the fit parameters a_i allows us to determine the time-dependent population $\rho_{11}(t)$ at physiological temperature $T = 300$ K as plotted in Fig. 5. Although we cannot discuss the energy transfer within the FMO complex over the 800 fs, we observe that coherent dynamics is much more pronounced in the CF-model compared to the other models investigated. At room temperature coherence is observed for more than 800 fs.

5. Conclusion

In this work, we present numerically exact results for the electronic energy transfer dynamics based on various models for the Fenna-Matthews-Olson (FMO) molecular aggregate. We determine its single excitation subspace dynamics within an open quantum dynamics approach based on dipolar couplings, site energies and their respective fluctuations from four different models including various degrees of detail, i.e. in the considered environmental fluctuations.

Table 1. Coherence times in [fs] in the various models for $T = 300$ K and $T = 77$ K.

	CF	IF	AR	OK
300 K	800	400	250	0
77 K		800	500	0

The Cho-Fleming model [21] and the Ishizaki-Fleming model [22] both use an Ohmic phenomenological environmental fluctuation spectrum to characterise the site energy fluctuations which orginate from solvent fluctuations as well as molecular vibrations. The Adolphs-Renger model [19] adds a single discrete vibrational mode to model the dominant molecular vibration following a detailed analysis of environmental fluctuations by Wendling et al. [20]. All three models were previously used to describe successfully experimental optical data. Nevertheless all three spectra already differ substantially. The Olbrich-Kleinekathöfer model [23, 24] is determined by extensive classical molecular dynamical simulations. It gives different spectra for each site and each spectrum consists of two continuous contributions resulting from solvent fluctuations and up to 13 discrete vibrational modes. Although these spectra are analyzed to fit the optical absorption spectrum of FMO as the other models the OK-model shows much stronger fluctuations than the others.

Investigating the population dynamics in all models we find coherent oscillations for site populations of pigments 1 and 2 in the CF-, IF- and AR-model. In the OK-model we find solely overdamped dynamics. The dynamcis of $\rho_{11}(t)$ for all four models in comparisson is plotted in Fig. 5. Coherence times are given in Table 1 and are decreasing shifting from the CF-, IF-, AR- to the OK-model. This could be partially expected from a simple one-phonon picture despite the strong non-Markovian charakter of the environmental fluctuations. Experimentally quantum coherence beatings in two-dimensional optical femtosecond spectroscopy were observed on time scales of more than 660 fs at 77 K [3] and about 300 fs at physiological temperatures [4]. Thus we must finally conclude at the present stage that currently the theoretical models can neither support the interpretation of the quantum beatings in the two-dimensional electronic spectra as purely electronic coherence nor can they rule it out.

Acknowledgments
We want to thank U. Kleinekathöfer and Valentyn I. Prokhorenko for fruitful and interesting discussions.

References
[1] van Amerongen H, Valkunas L and van Grondelle R 2000 *Photosynthetic Excitons* (Singapore: World Scientific)
[2] May V and Kühn O 2004 *Charge and Energy Transfer Dynamics in Molecular Systems* 2nd ed. (Weinheim: Wiley-VCH)
[3] Engel G S, Calhoun T R, Read E L, Ahn T K, Mančal T, Cheng Y-C, Blankenship R E and Fleming G R 2007 *Nature* **446** 782
[4] Panitchayangkoon G, Hayes D, Fransted K A, Caram J R, Harel E, Wen J, Blankenship R E and Engel G S 2010 *Proc. Natl. Acad. Sci. USA* **107** 12766
[5] Fenna R E and Matthews B W 1975 *Nature* **258** 573
[6] Milder M T W, Brüggemann B, van Grondelle R and Herek J L 2010 *Photosynth. Res.* **104** 257
[7] Collini E, Wong C Y, Wilk K E, Curmi P M G, Brumer P and Scholes G D 2010 *Nature* **463** 644
[8] Reineker P 1982 *Exciton Dynamics in Molecular Crystals and Aggregates* Springer Tracts in Modern Physics, vol. 94 (Berlin: Springer)

[9] Ishizaki A and Fleming G R 2009 *J. Chem. Phys.* **130** 234110
[10] Nalbach P and Thorwart M 2010 *J. Chem. Phys.* **132** 194111
[11] Mohseni M, Rebentrost P, Lloyd S and Aspuru-Guzik A 2008 *J. Chem. Phys.* **129** 174106
[12] Plenio M B and Huelga S F 2008 *New J. Phys.* **10** 113019
[13] Olaya-Castro A, Lee C F, Fassioli F and Johnson N F 2008 *Phys. Rev. B* **78** 085115
[14] Nalbach P, Ishizaki A, Fleming G R and Thorwart M 2011 *New J. Phys.* **13** 063040
[15] Thorwart M, Eckel J, Reina J H, Nalbach P and Weiss S 2009 *Chem. Phys. Lett.* **478** 234
[16] Ishizaki A and Fleming G R 2009 *J. Chem. Phys.* **130** 234111
[17] Nalbach P, Eckel J and Thorwart M 2010 *New J. Phys.* **12** 065043
[18] Chen X and Silbey R J 2010 *J. Chem. Phys.* **132** 204503
[19] Adolphs J and Renger T 2006 *Biophys. J.* **91** 2778
[20] Wendling M, Pullerits T, Przyjalgowski M A, Vulto S I E, Aartsma T J, van Grondelle R and van Amerongen H 2000 *J. Phys. Chem. B* **104** 5825
[21] Cho M, Vaswani H M, Brixner T, Stenger J and Fleming G R 2005 *J. Phys. Chem. B* **109** 10542-56
[22] Ishizaki A and Fleming G R 2009 *Proc. Natl. Acad. Sci.* **106** 17255
[23] Olbrich C, Jansen T L C, Liebers J, Aghtar M, Strümpfer J, Schulten K, Knoester J and Kleinekathöfer U 2011 *J. Chem. Phys. B* **115** 8609-21
[24] Olbrich C, Strümpfer J, Schulten K and Kleinekathöfer U 2011 *J. Chem. Phys. Lett.* **2** 1771-76
[25] Nalbach P, Braun D and Thorwart M 2011 *Preprint arXiv:1104.2031v1*
[26] Tronrud D E, Wen J, Gay L and Blankenship R E 2009 *Photosynth. Res.* **100** 79-87
[27] Makri N and Makarov D E 1995 *J. Chem. Phys.* **102** 4600
[28] Makri N 1995 *J. Math. Phys.* **36** 2430
[29] Nalbach P and Thorwart M 2009 *Phys. Rev. Lett.* **103** 220401
[30] Nalbach P and Thorwart M 2010 *Phys. Rev. B* **81** 054308
[31] Wen J, Zhang H, Gross M L and Blankenship R E 2009 *Proc. Natl. Acad. Sci.* **106** 6134